PLAYFUL
MATHEMATICS

PLAYFUL
MATHEMATICS

for children 3 to 7

HELEN J. WILLIAMS

CORWIN

A SAGE company
2455 Teller Road
Thousand Oaks, California 91320
(0800) 233-9936
www.corwin.com

SAGE Publications Ltd
1 Oliver's Yard
55 City Road
London EC1Y 1SP

SAGE Publications India Pvt Ltd
B 1/I 1 Mohan Cooperative Industrial Area
Mathura Road
New Delhi 110 044

SAGE Publications Asia-Pacific Pte Ltd
3 Church Street
#10-04 Samsung Hub
Singapore 049483

Editor: Amy Thornton
Senior project editor: Chris Marke
Project management: River Editorial
Marketing manager: Dilhara Attygalle
Cover design: Wendy Scott
Typeset by: C&M Digitals (P) Ltd, Chennai, India

Library of Congress Control Number 2021944697

British Library Cataloguing in Publication Data

A catalogue record for this book is available from the British Library.

ISBN 978-1-5297-5515-2 (pbk)
ISBN 978-1-5297-5516-9

CONTENTS

ABOUT THE AUTHOR

Helen J. Williams has worked for many years in education, teaching from Reception into Key Stage 2. Helen's particular expertise is in Early Years, Key Stage 1 and early Key Stage 2 with a special interest in developing effective, playful opportunities for learning mathematics. As well as writing, Helen runs Early Years courses and contributes to conferences. Her doctoral research was completed with Roehampton University in 2014 and is entitled *The Relevance of Role Play to the Learning of Mathematics in the Primary Classroom.*

She is a member of the British Society for Research into Learning Mathematics (https://bsrlm.org.uk), the Association of Teachers of Mathematics (www.atm.org.uk) and a founder member of both the Early Childhood Mathematics Group (https://earlymaths.org) and the Firm Foundations group of early educators (https://ffed.weebly.com).

She tweets as @helenjwc and blogs (sporadically) here: https://info125328.wixsite.com/website

ACKNOWLEDGEMENTS

I learned it because I allowed my students to enlighten me (Caleb Gattegno 1988).

With my thanks to all the many children and colleagues I have had the privilege to learn from. And most of all very special thanks to my family and friends who have always spurred me on.

PREFACE

I am an independent educational consultant specialising in the learning and teaching of primary mathematics. My expertise is in Early Years and KS1 mathematics and that is where I have focussed this book. I have worked (and played) in mathematics education for over 30 years and am honoured to call myself a member of some eminent mathematics associations. However, I managed to get right through school, university and my PGCE without mathematics being either significant or enjoyable to me. Reflecting on my journey to becoming passionate about the learning and teaching of mathematics, I can spot my starting point.

My first brush with mathematics I found enjoyable was as a new teacher with my very first class of thirty-five 7–8-year-olds, in 1979. Together we watched the ITV series *Leapfrogs* once a week on an old television set in the hall. *Leapfrogs* was a series of twenty-eight, 15-minute 'magazine' programmes for the 7–9 age range aimed at stimulating mathematical activity in the classroom. I still have my treasured and battered copy of the teachers' handbook (I have written my name and 'lay off!' inside the cover) where the introduction refers to children creating mathematics and experiencing what it feels like to 'act like a mathematician'. It wasn't only the children that experienced this in that 1979 classroom and I eventually discovered that the architects of *Leapfrogs* were those I have since been delighted to call colleagues in the Association of Teachers of Mathematics (https://www.atm.org.uk).

Watching the visual images and curiosities presented in these ground-breaking programmes and following up the suggestions of how I might develop these ideas with my first class, was my introduction to being mathematically curious and exploratory, it was the first time I cared about, sought an answer to or asked 'why'? And I loved it.

This book is a collection of what I have been working on, reading about, researching and reflecting upon with children and colleagues since that time.

Some might now refer to such visual and pro-active stimulation as 'provocations'. I hope this book is a provocation to you, the reader. An invitation to enjoy mathematics alongside your children.

I am contactable on Twitter as @helenjwc, where you will find my blog. Stay in touch, it would be great to hear what comes of me writing this book.

1

WHY PLAYFUL MATHEMATICS?

Mathematics could be like roller-skating, but usually it is like being told to stop roller-skating and come in and tidy your room.

This is not a superficial matter.

(Winter 1992: 99)

INTRODUCTION

This book aims to inspire us to re-think mathematics, to explore how children learn best and how adults can best support their mathematical learning. In the coming chapters I explore what mathematics is and might be for young learners. Much of what is said is applicable to learners of all ages, but primarily this is a book written to support Early Years educators in developing their teaching of mathematics for children between the ages of 3 and 7. Secondly, it is a book to inform senior leaders of the particular issues in Early Years teaching and learning and to stimulate thinking and discussion around mathematics, school-wide.

I have based this book in research evidence, in my own practice and in that of many inspirational colleagues. Importantly, it is based also in what children have taught me as I have had the privilege to work with them over the last 30-plus years. Every interaction and vignette included within this book has taken place.

WHAT DO I MEAN BY 'EARLY YEARS'?

Internationally, Early Years as a developmental phase is acknowledged as lasting from birth to 8 years of age (Downton et al. 2020). England and Australia are out of step with most of Europe in expecting children to start compulsory schooling at 4 and 5 years old (Sharp 2002) as in most of Europe, and New Zealand, children begin compulsory primary schooling when they are 6 or 7 years old (Education, Audiovisual and Culture Executive Agency 2018). The English school starting age has remained unchanged since it became compulsory in the 1870 Education Act and has engendered hot debate over many years. Its bearing on this book is two-fold. Firstly, much of the research I examine has been undertaken outside the UK, where children are often educated in a state-funded kindergarten system and begin formal school later. Secondly, a 5-year-old is a 5-year-old whether they are in a Reception class in England, a pre-school in Scotland or a kindergarten in Sweden. What this book aims to do is to examine the research that has important implications for the mathematical teaching and learning of a 5- (or 3-, 4- or 6-) year-old, wherever they are educated. This means that we will be questioning the appropriateness of some of our school environments for young children's mathematical development.

TEACHER, PRACTITIONER OR EDUCATOR?

I write this book for all adults who work with children between the ages of 3 and 7. This includes families and carers. I will be using the terms 'practitioner' and 'educator' in this book rather than 'teacher' because many of the adults who work with children under the age of 5 do not have qualified teacher status (QTS) but do all the things that 'teachers' do:

> [Teaching] is a broad term that covers the many different ways in which adults help young children learn. It includes their interactions with children during planned and child-initiated play and activities, communicating and modelling language, showing, explaining, demonstrating, exploring ideas, encouraging, questioning, recalling, providing a narrative for what they are doing, facilitating and setting challenges.
>
> (OFSTED 2019a :33, 2019b: 80)

'Educator' comes from the Latin 'to lead out'. This book is for *all* of us, wherever and who-ever we are, working and interacting with children between the ages of 3 and 7.

Figure 1.1

Figure 1.2

The idea that babies and very young children are blank slates has been completely debunked by research over the last 30 years. The notion that babies were irrational, amoral, unable to understand relationships between cause and effect, or see things from different points of view, was reached without much evidence:

> *That view has been completely turned around. Every week we discover that even the very youngest babies know more and learn more than we ever could have conceived was possible before.*

(Gopnick 2015)

Alison Gopnick's research is evidence that if we ask the very young questions in languages they understand, they can answer and inform us. By languages we do not mean 'foreign' languages, but tuning in to their ways of expressing themselves. Malaguzzi has referred to this as the 'Hundred Languages of Children' (Malaguzzi 1993; Nursery World 2007), and Donaldson (1978) as situations that 'make human sense'. This book does not specifically refer to working with babies and toddlers up to the age of 3, nor to young learners older than 7, although much of what is said applies to these learners. The examples I draw on lie in the 3–7 age band. Birth to 2 years is a specific phase requiring its own book. There are a number of websites that support the mathematical development of young children including babies, and I recommend the following:

- The Early Math Collective at the Erikson Institute, Chicago: **https://earlymath.erikson.edu**
- The UK-based Early Childhood Mathematics Group: **https://earlymaths.org**
- Development and Research in Early Math Education at Stanford. **https://dreme. stanford.edu**

WHAT IS IT ABOUT MATHEMATICS?

What is it that makes many of us feel inadequate and quick to admit we could never 'do mathematics'? The answer lies in the way it is taught. Mathematics teaching does not have a good track record for inspiring enjoyment and creativity, let alone love, amongst much of the population. My own experience of mathematics at school was not enjoyable. I learned to enjoy it as I taught my first Reception class in the 1980s. I learned to enjoy it as I observed and pondered upon my Reception children's responses to what I was offering them mathematically and as I discussed this with, and worked mathematically alongside, other professionals. Most of my professional life has been spent working with children and adults to widen both views of mathematics and how it can feel to engage in it. To do so, we have engaged in playful inquiry, an example of which follows.

Take a moment to read the vignette 'Winning the Goldest', now. What do you *notice* about this moment I recorded in a Nursery setting? What do you *wonder*?

When I recorded this, I was struck at the playfulness of the moment. Playfulness does not happen by chance. It has to be prepared for. Small decisions make big differences. What have I done to encourage this as a playful activity? I have provided resources that I think might interest these children, in this case, a large heap of real 1p coins, previously soaked in a well-known fizzy drink to make them shine, small containers with lids to hold them (not too many for them to count) and dice, to imply a game. I don't have to say anything. In fact, it is often important that I do not say anything, but watch and listen. A game begins, of collecting coins according to the dice-roll, seeing how many coins they can fit into their pots and discussion about who might have the most coins. There are many mathematical opportunities to follow up, plus one I had not foreseen; what different numbers of coins 'sound like' when pots are shaken. A game is invented where one child shakes their pot of coins and others have to guess how many are inside. Lots of counting and predicting ensued. Then the delightful contradictory moment arose when a pot stuffed full of coins was shaken and made no noise, surprising even the instigator. *Ha!* he said as he tipped out his coins. A mathematical joke we can all enjoy: we all thought it was empty.

VIGNETTE: WINNING THE GOLDEST

Figure 1.3

A group of Nursery children aged between 3 and 4 years of age are sat with a pile of 1p coins, some small pots with lids and dice.

Josh rolls a die, counts the dots and counts the corresponding number of coins into his pot.

Chantelle:	Are you putting them all away?
Josh:	No.
Freya:	I've got all three, it depends on how many you got. One, two, three (*laughs*). Three! Again, again, again, again!
Josh:	I want a turn.
Freya:	I got three.
Sacha:	I got four.
Sacha:	Let's shake our pots again, yeah?
Josh:	Where's my lid?

Everyone starts filling their pots with pennies and shaking them.

Josh:	(*several times, loudly*) I want a turn!
Sacha:	I got most of the money, 'cos I winned.
Freya:	You haven't winned the goldest. You have to have many.
Sacha:	I have all of them.

Figure 1.4

WHAT DO I MEAN BY PLAYFUL?

Although the relationship between play, learning and development has been extensively studied, 'play' seems to be a difficult concept to define. One possible explanation for this is that whether or not something is considered play is down to how the situation is perceived by the individuals involved (van Oers 2013). Do our children, when we are playing a dice game, for example, see the game as play? Are they curious, motivated and creative? These are the factors that drive learning. Or do they immediately move away when I move to work with another child or task? How can I present something in order for a child to become deeply engrossed?

What is 'playfulness'? Being playful can be described as having a disposition and a willingness to engage in play (Education Scotland 2020). In the vignette above, although I have provided the materials with some mathematics in mind (counting), I am open to what I might observe. I am being playful so that the children might also be so. I might ask: *What can we do with these?* I might just stay silent to begin with. Playfulness is the difference between lining up some pennies and asking children to count them and having a heap of pennies and saying: *I wonder, how might we count these pennies?* It is the difference between teaching children the rules of a game and inviting them to invent a game.

Figure 1.5

This does not mean we never teach them the rules of a game, it is about balance. Research is clear on the need to foster children's curiosity (Cheeseman and McDonough 2016; Downton et al. 2020) and playfulness nurtures this:

> *The implications of this research for classroom teachers were threefold: to offer children challenging tasks and interesting mathematical tools: listen carefully and be aware of the learning potential children bring to the task; and encourage curiosity by showing a real interest in children's investigations.*
>
> (Downton et al. 2020: 151)

Mathematics teaching must allow space for children's sense-making as well as their own ideas. This is teaching where my – often too many – prepared questions do not come too fast or too soon. Teaching that allows space for children to think and to engage in their own playful inquiry. Teaching that allows gaps for playing with ideas, maybe on paper, by thinking alone, by talking, and mainly by manipulating – playing with – the materials. Teaching that allows space for me to observe, and thus notice, possible next steps and links. To build on where

Figure 1.6

the children are coming from and what they know. This is mathematics that contains an essential ingredient of 'healthy work':

> *Healthy work – that is, work which is seen by the participant as enjoyable, purposeful, creative and rewarding – [has] the essential ingredient of play: the facility for becoming absorbed, for losing oneself in the activity.*

(Brown 1996: 113)

Figure 1.7

SOME HISTORICAL CONTEXT

4-year-old Sam to adult:	What's a hundred and a hundred?
Adult:	Two hundred.
Sam:	What's a hundred and a hundred and a hundred and a hundred? I think it's probably four hundred. And a hundred and a hundred and a hundred and a hundred and a hundred, that's five hundred.
4-year-old Alice to Sam, also 4:	Can you count how many spots are on my dress?
Sam:	Lots.
Alice:	A hundred, I 'spect it's a hundred. Until my birthday we'd be counting.

In my experience, and through no fault of our own, Early Years professionals can struggle with subject knowledge and making mathematics relevant and engaging for our young learners. In their play, children are naturally mathematical; they make patterns, attend to symmetry, show interest in variations in quantities and weights, are fascinated by time. Yet practitioners are often unsure about how to build on this or how this observed play 'fits' in with the mathematics they are required to teach. We are unclear how much to direct what happens and how much to leave the activity to play out, what Shulman (1986) refers to as *pedagogical subject knowledge*. Those of us who have had negative mathematics experiences are anxious about how we teach mathematics without passing this lack of confidence on to our learners. Good Early Years mathematics includes both teaching and the space and freedom for young children to explore their and others' ideas, to predict, to reason, to explain, to wonder. But often these opportunities are missed in mathematics, where there is a tendency to overcorrect, to steer children closely through a series of small, pre-determined steps, to carefully avoid the making of mistakes, the challenging, the pure, wild enjoyment. In fact, we need to apply what we know is critical in other curriculum areas to mathematics. In the words of Winter (1992), we need to roller-skate rather than just tidy our rooms.

Research over time has pointed to learners of all ages and attainment levels, and in all jurisdictions, as holding negative attitudes to mathematics (Askew et al. 2010). Females are more likely to pursue higher education than males (Department for Education 2007), but STEM subjects (science, technology, engineering and mathematics) remain an area in which women are significantly underrepresented and there is a persistent trend of an over-representation of males in those considered as 'high achievers' in both mathematics and science (Meinck and Brese 2019). The situation is complicated by the fact that England, Scotland, Wales and Northern Ireland are four countries that do not require compulsory participation in mathematics after the age of 16 (Hogden et al. 2010). Girls do not pursue mathematics as commonly as boys for many complex reasons, maybe because they consider they have other options, so this may be related in part to lack of pleasure or interest.

Moreover, in a classroom focused on correct answers at the expense of anything else, the safe option might be to stop taking a risk, to stop engaging in harder challenges; to stop playing mathematically. Even those viewed as 'good at maths' have a lot to lose if they take a risk, as they might get something wrong and lose status. At any age, over-emphasising the fast answer or the 'correct' procedure to the expense of all else can cause children to become obsessed with not 'doing something the right way' and ultimately fearful of maths. We certainly know that children who believe they are naturally 'no good' at mathematics are less successful than those who believe they can learn. If we group children by 'ability', we effectively tell some of them that they are not as good at mathematics (Boaler 2013; Gifford 2015).

Over the last 10 years research has moved firmly into the foreground of educational discussions. Sadly, as is often the case with research, much of it is cherry-picked for political ends, or so far adrift from the classrooms or settings it purports to describe to be virtually useless. All the research explored in this book has been selected because it has been undertaken with children between the ages of 3 and 7. In recent years we have learned a lot about how mathematics is learned in the Early Years; however this is not often reflected in the politically driven, damaging downward pressure from those in positions of power, or in the advice of well-intentioned, but ill-informed, colleagues. This book is an attempt to counter this by interpreting relevant research practically. I will not be referring to research with older students and watering this down for the Early Years.

Downward pressure on the Early Years Foundation Stage (EYFS), often from account-ability requirements, often from much later in school, such as statutory testing and requests for data to demonstrate progress, impinges on what is provided, especially in Reception classes. In particular, sections of the Early Learning Goals (statutory targets for children to achieve at the end of their Reception year in England) that are developmentally inappropriate and unevidenced by research are detrimental to effective mathematics teaching, as practitioners are driven to realise these, despite the fact they might not be achievable by this age of child. As educators we need to educate ourselves about this evidence and become coherent about articulating this, as well as why we do what we do, to our colleagues. I hope this book helps spread our message that early childhood mathematics is a well-evidenced phase, with an accompanying Early Years pedagogy crucial for developing children's mathematical well-being for future learning.

SAFETY

What does it mean – learning in a place of safety? I myself did not have a very positive secondary school maths experience and I came late to enjoying mathematics; this has probably made me super-sensitive to the atmosphere that surrounds mathematics teaching. To change the atmosphere from one where the speedy, correct answer is valued above all others, e.g. *Ten! A triangle!*, I need to provide mathematics that is both varied in terms of content and woven through with a strong thread of exploration and play.

Figure 1.8

Being deliberately playful with mathematical ideas is critical in creating a safe environment for children to think and operate mathematically. If something is approached as a game, as an exploration, through pretence or via a story, children can try out ideas and make suggestions without fear of failing. Playful inquiry creates a place of safety for both child and adult to explore and enjoy some mathematics. Every interaction, with every child, should result in them thinking positively about themselves as mathematics learners. It should be safe for every child to speculate, to make a statement about what they notice,

Figure 1.9

to have a guess, to say something silly. I must make sure I don't simply jump on and reward the answer I am expecting, but remain open to others as a way into the child's understandings. A safe environment involves building a mathematical learning community where all views are respected and listened to, where everyone's contribution impacts on our cumulative learning.

Noddings (1984) argues that the teacher's task is to receive and accept students' feelings towards a subject as part of what we teach. She argues that achieving our ends instructionally, but leaving the student disliking the subject, means that we have failed educationally. Everything we do as educators has to be built around the positive relationships we have with the children, and mathematics must be no exception to this. The environment we provide is not just about physical provision and resources, but also about how inclusive we make our mathematical environment, how included in mathematics all our children feel. This means that even the unexpected response is received positively; for example, how might I respond positively to the child who says:

Child: I'm four. Four metres, that's my old.

How might I respond so that this child is keen to continue this conversation and to begin similar mathematical conversations, in order they do not become fearful of making a mistake and these conversations stop? Recognising and responding to what they *do* know is one way: *Yes, four years* **old** *is your age, and three years old is your sister's age*, for example. Or. *yes, you are four years old and I am 21 years old.*

Mathematics exists in and builds on the environment we provide and the relationships we foster with our learners. From a young age, children can either learn that mathematics is nothing to do with them, or that they can play an active, creative role in acting and thinking mathematically. For example, here are two nearly 4-year-olds:

(Sam holds up four fingers on each hand.)

Sam: How many is this, Alice? Try and guess.

(Alice uses her fingers to start to count aloud.)

Sam: No, you mustn't count. Try and guess. Try and *guess*. Don't count, you must *guess.*

Alice (*still counting*): It's eight.

Sam: You should have guessed.

Alice: I didn't know. (*Pause. Holds up four fingers on each hand.*) *Now* I will guess. It's *eight*, Sam!

To raise a generation of confident mathematical thinkers, we need to embrace mathematical playfulness. Being playful in mathematics is about opening up what we offer to children so that more than one solution is possible and where there is plenty of space for them to choose both to tackle a task in different ways and to extend the task into areas we had not planned for. Children need the time and space to play with concepts. This book contains examples of playful tasks started by adults and by children; it explores how we might respond to and build on these. Playful mathematics can and should be challenging. By challenging I mean a task that makes children think, rather than notions of increasing difficulty. Johnson and his colleagues found that many pre-schoolers demonstrated more sophisticated understandings of counting whilst engaged in tasks requiring extra challenge, that were not captured by simpler tasks. They found that children showed more counting ability when asked to count 30 pennies than 15, although 30 was seen as beyond their capabilities (Johnson et al. 2019). This seems to indicate that it is possible to underestimate young children's abilities and that if the task is interesting enough, children will rise to the challenge. Russo (2015) looked at challenging tasks and how pupils approached them. He found that pupils of all ages enjoyed being challenged. One example of a challenging task he used with 6- and 7-year-olds was this:

Without leaving your seat, or talking to anyone, can you work out how many feet are in the room right now? Show how you worked it out.

(Russo 2015: 2)

This book is aimed at those who teach mathematics to our youngest children, its challenges and its delights. Every example of children thinking and working contained in this book took place and was recorded by me – whether as notes, video or voice recording. Far too often mathematics is about tidying our rooms instead of roller-skating. So, let's get stuck in and roller-skate together.

All successful education, I would argue, aspires to the conditions of play.

(Winter 1992: 82)

The following chapter, 'Fostering mathematical reasoning', examines the research behind reasoning and problem solving in Early Years mathematics and how we might act to support children in developing these areas and to build mathematically on what they know and can do.

Figure 1.10

2

FOSTERING MATHEMATICAL REASONING

Reasoning is fundamental to knowing and doing mathematics ... Some would call it systematic thinking. Reasoning enables children to make use of all their other mathematical skills and so reasoning could be thought of as the 'glue' which helps mathematics make sense.

(NRICH Primary Team 2014)

WHAT IS MATHEMATICAL REASONING AND WHY IS IT IMPORTANT?

Reasoning is logical, purposeful thinking (Borthwick and Cross 2018). Mathematical reasoning is related to problem solving, whilst problem solving involves the 'how' of tackling a task, selecting, applying and evaluating; reasoning works on the *'why'*. Mathematical reasoning is the critical skill that enables us to make use of all other mathematical skills. Reasoning is important because it is the heart of mathematics: seeking to make sense and to be understood, reflecting on patterns you notice, solutions to problems and whether or not they are reasonable. Reasoning has to involve some logical thinking: *if ... then ... because* and also, eventually, a move towards generalisation and explanation; by *justifying, convincing* and *proving* findings. Mason and his colleagues (2012) point out that, whilst a justification may be adequate in one community, it might not in another, and puts forward a sequence of justification, or proof, as: *convince yourself, convince a friend, convince a sceptic* (Mason et al. 1982). *Convincing a sceptic* being the ultimate proof, involving an attempt to satisfy *all possible* objections with a widely acceptable justification. Of course, this is very high-level thinking, and not to be expected of young learners, but what we can do is to begin the process of reasoning by:

- *getting a sense of* justification and proof, as a preliminary stage to justifying and proving;
- nurturing *convince yourself* as a growing sense of *why* a conjecture is correct, for example, by regularly asking children, *Are you sure?*;
- beginning to ask more experienced children to *convince a friend* about something they have noticed, for example by asking, *Can you explain to Rudi* why *you think that?*

Most children between 4 and 7 years of age can begin to get a sense of what 'convince' means by us using and modelling its use. Even if we do not receive any immediate reply to questions such as these, the first stage is for children to become familiar with hearing them. Very young children do reason about the world around them, noticing patterns and relationships. Many children ask 'why?' over and over in a search for some firm foundations in what they are experiencing and noticing. Often, I find that mathematical 'whys' occur in the form of an announcement, for example, a toddler exclaiming *Dog!* when seeing a dog that is unlike any dog they have previously seen. Here, with the youngest children, reasoning is often implicit. One mathematical example is a 3-year-old repeatedly turning over a 'Snakes and Ladders' game board, maybe to see if there are numbers (continuing?) on the reverse; or the 4-year-old here:

Child: What does five and five make?

Adult: Ten. (*Holds up two hands of five fingers.*) Five and five, ten all together.

Child: No! No, *no*. One and nought make ten so what do five and five make?

Here 'make' is understood by the adult as 'combine' whereas the 4-year-old is reasoning about and asking how numbers are written. Reasoning does not have to be verbal, and often it is not with the youngest children, or children with special educational needs.

Figure 2.1

Watching a toddler with a 'shape-posting' toy, we can see the visual links they are able to make between the blocks and the holes. By doing this, they are reasoning *spatially*, although they are not able to verbalise this: *If I turn it like that it will go in there.* The experienced toddler will often select the cylinder to post first; maybe they 'know' this is the easiest of the selection to post. Goswami's research points to young children thinking and reasoning in largely the same ways as adults; the difference being in their lack of experience (Goswami 2015). Mathematical reasoning, even more so than children's knowledge of arithmetic, contributes positively to children's later, wider, achievement in mathematics (Nunes et al. 2009, 2015). Back (2004) suggests that classrooms in which teachers encourage pupils to participate in mathematical reasoning and argument are more likely to be successful in helping children to be enthusiastic mathematicians, and that, after all, is what we are pursuing.

SOME EXAMPLES OF YOUNG CHILDREN REASONING MATHEMATICALLY

VIGNETTE: OBSERVING A 3-YEAR-OLD WITH BLOCKS

Figure 2.2

Here is Bao (3 years of age) choosing blocks to place on the top of the platform. It is a physical challenge to carry each one up on to the platform every time. It is not a random selection of blocks. He chooses them and fits them together with care, end to end to stretch across the platform. There is a lot of movement all around him, but he is concentrating hard on getting his blocks aligned. We can observe that he is reasoning about the shape and length of those that will 'work'. He then builds the wall up, choosing the blocks to balance well. He builds an upright structure and manages an arch with a vertical internal space.

A few older children join in at a few points, choosing, moving and placing the blocks and Bao is happy with that.

Whilst Bao is unable to explain what he is doing in English, as he has yet to acquire the language to do so, he is clearly seeking to make sense of the blocks and how to use them, to realise a structure he has in his mind; he is enacting: *if ... then ... because*. He is also working at being understood by other children, reflecting on his solution to the task he has set himself, by adjusting what he is doing and accepting (or otherwise) what other children are doing and suggesting.

Figure 2.3

Block and pattern play are perfect places to observe very young children reasoning. If I see a walled enclosure with a doorway, it is clear that the children have made decisions about selecting blocks that match in height and shape in order for the walls and doorway to function. Casey et al. (2008) spell out a useful trajectory for block play, from random block placement through horizontal enclosures one block high, to enclosures of height with roofs, entrances and divided internal spaces. At all these stages, we can observe from what they do and make, children making decisions based on internal reasoning: *if ... then ... because ...* If we make ourselves aware of this, and as they become more experienced, we can support children to explain their choices and, with sensitive questioning, justify the decisions they make; to convince themselves, to convince a friend and maybe to convince a sceptic.

VIGNETTE: 'WHO HAS MOST?'

A few 4- and 5-year-olds are challenged to find out which of two toys 'has the most money' and to show their solution on paper.

Joanne lays her coins out in a horizontal line, and counts them aloud, touching each one from left to right. She writes each numeral from one to ten above each coin, occasionally asking another child a question: *What does a seven look like?*

As she finishes, she moves the coins from the paper and says aloud:

I've got ten here. This is more, look.

Figure 2.4

She explains to the adult: *I have drawn a line so you can see this is for the dog and this is for the bear. There is more for the dog, because one more here and one more here, is missing,* and she touches her paper twice after the numeral eight.

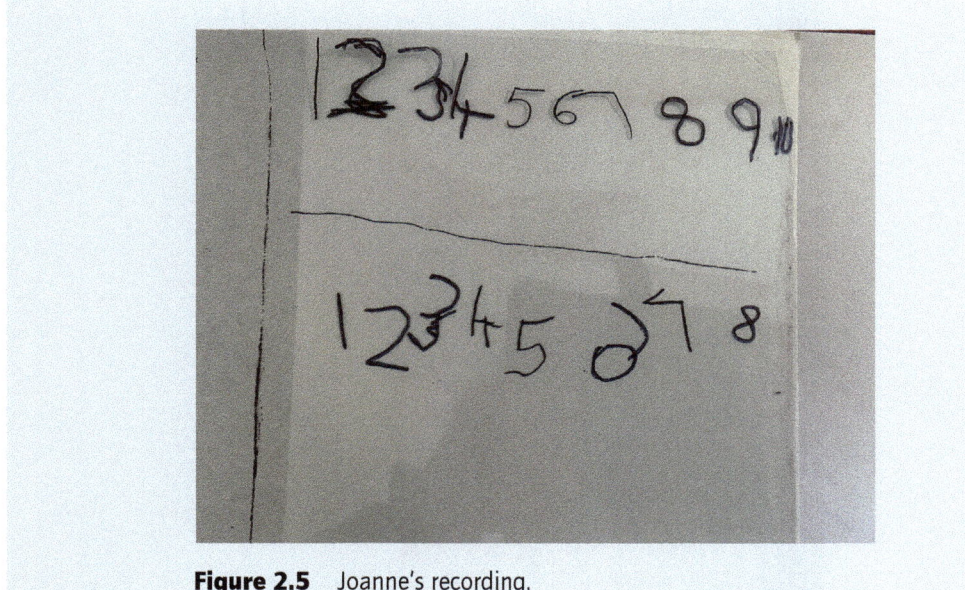

Figure 2.5   Joanne's recording.

Joanne solves this 'who has more?' problem efficiently, proving the dog has more coins, by lining up the two sets of coins alongside one another. It is significant that she was asked to 'prove it' on paper; this has supported her reasoning. Here she uses her recording to explain her reasoning to the adult. By choice, she uses her knowledge of the numeral sequence to illustrate how many coins each toy has, and then justifies who has more, by referring to her recording and saying: because *one more here and one more here is missing*. In other words, the bear does not have the 'ninth' and the 'tenth' that the dog has. When faced with a mathematical challenge, it is reasoning that helps us to make use of context and relevant prior knowledge: Joanne makes use of both in this example.

Positively encouraging children to record their own ways of representing their mathematics *as they are doing and thinking about it*, rather than afterwards, is important for linking the concrete, the pictorial and the abstract, and one way we can 'see' children's reasoning (Davenall 2016).

Overleaf we can see Rose (5 years of age) using drawing to think through that ten more children than 105 would be 115, and Alfie, who is in Nursery, unpicking the number 17 (Figure 2.7).

In Figure 2.8 another 5-year-old child keeps track on paper of how they positioned eight toy ducks between pond, decking and nest.

Figure 2.6   Rose counts on 10 from 105.

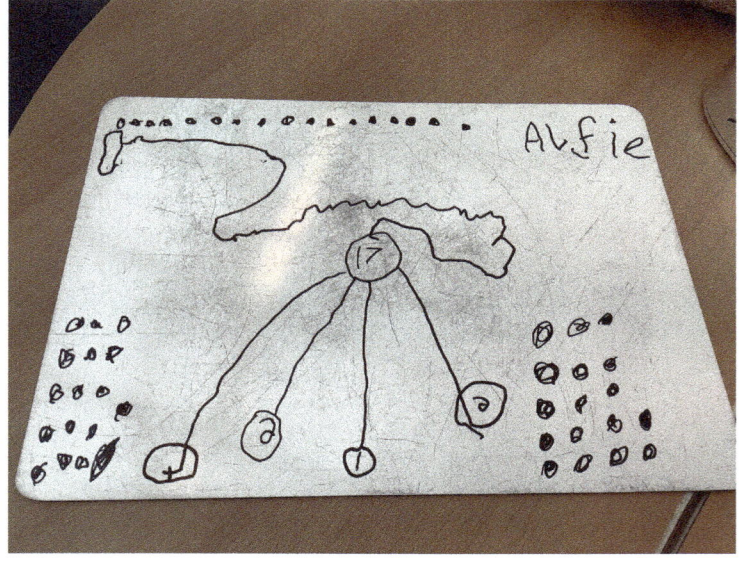

Figure 2.7   Alfie unpicks 17.

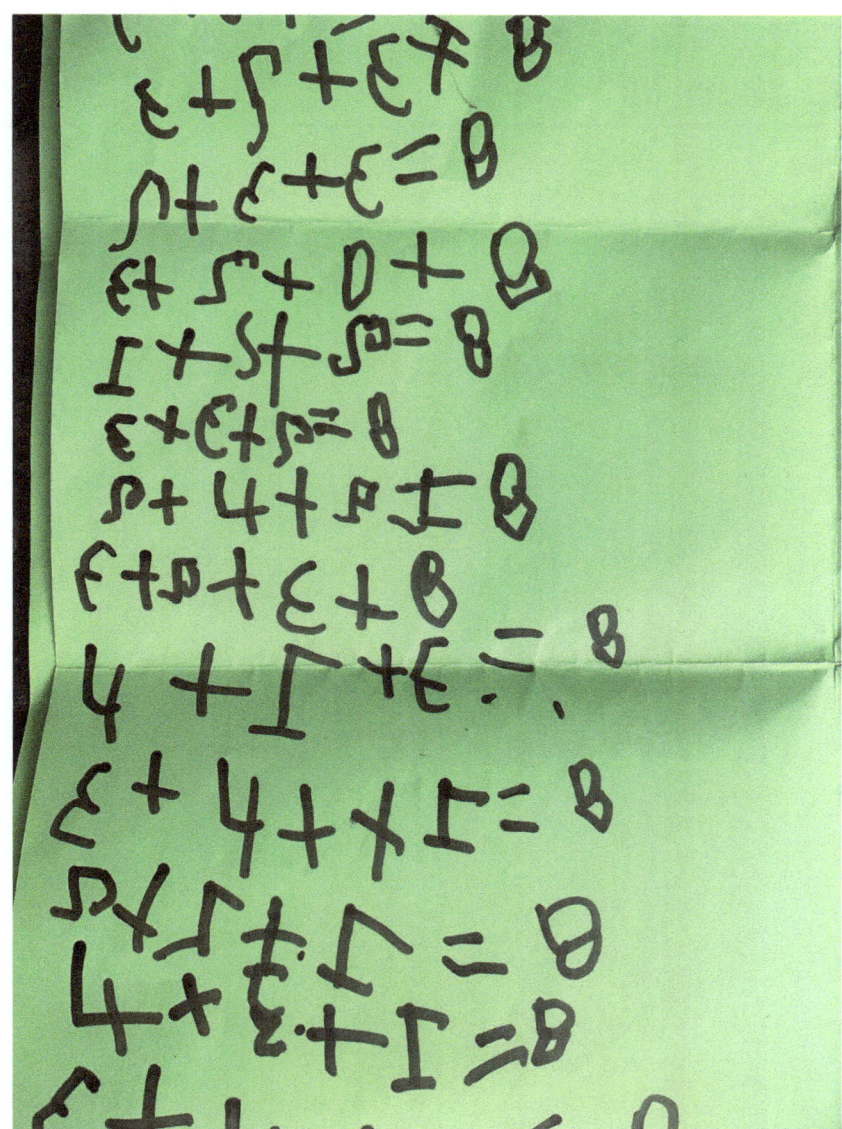

Figure 2.8   Recording 8 ducks positioned in 3 different places

Presenting young children with problems that engage them and that encompass *prediction* can lead into reasoning and this chapter examines a range of examples for supporting this.

USING ERROR-SPOTTING, GAMES AND STORY PROBLEMS TO STIMULATE REASONING

ERROR-SPOTTING

VIGNETTE: COUNTING OVER 100

Figure 2.9

Emil (5 years of age) is with a group of children watching toy Milo 'count' above 100.

Adult: Milo is going to start counting on from 100; watch him carefully. Look …

Child:	He might miss something out.
Adult as Milo, who touches the counters as he counts:	One-hundred-and-one, one-hundred-and-one, one-hundred-and-two, one-hundred-and-three ... (*interrupted*)
Emil:	Yes (*pause*). No! There would need to be one-hundred-and-one and then one-hundred-and-two.
Milo:	Ah! One-hundred-and-one, one-hundred-and-two, one-hundred-and-two, one-hundred-and-three.
Emil:	He's getting it wrong again!
Adult:	So – what does he have to do?
Child:	He's not even talking!
Emil:	Umm, one-hundred-and-one, it is then one-hundred-and-*two*, one-hundred-and-*three*.
Adult:	Ahh!
Child:	Go him!

Setting up an error-spotting situation is an effective way of refocusing children from the doing of an activity to thinking and reasoning about it. Rather than their attention being on the action of counting – thinking about what to say next, whether they have already moved their finger, exactly when to touch the next one, and so on – they can reflect upon counting, the *how* and explaining *why* it is correct or *why not*. In the vignette above, although the children are clear that the toy is not actually counting, Emil discerns the error and corrects it. He also applies his knowledge of the counting sequence, and by adding *it is then* explains his correction that the counting sequence after 100 has to start again from one. He is 'convincing himself' and attempting to 'convince a friend' (Mason et al. 1982).

The work of Thouless and her colleagues (2020) and of Borthwick and her colleagues (2021) are powerful examples of how error-spotting can be used to support mathematical reasoning. Thouless and Borthwick built on the work of Papic and her colleagues (Papic et al. 2015) researching 3–5-year-old children's abilities to recognise, continue, copy and create repeating patterns using a wide range of loose parts. Children had been immersed in exploratory pattern-making and sensitively challenged to develop their linear pattern-making to create a repeating border pattern 'that worked' around a plate or closed shape. This was far more challenging mathematically, as the pattern, rather than continuing infinitely, had to complete in a set distance, around a border, and meet up satisfactorily.

Figure 2.10

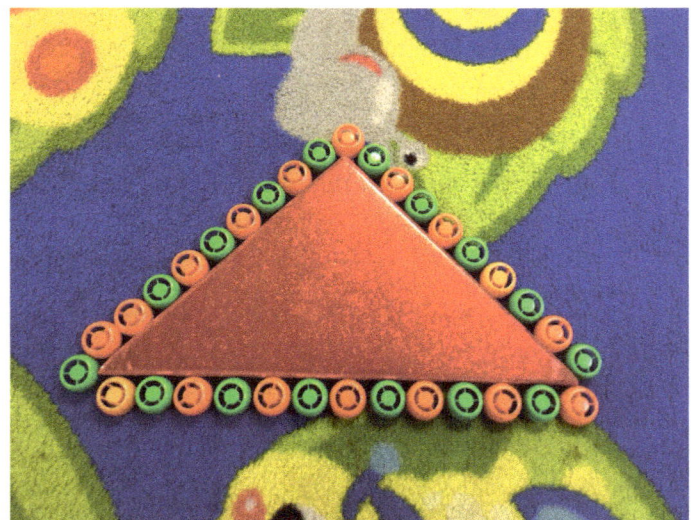

Figure 2.11

Source: Borthwick, Gifford and Thouless (2021)

Continuous patterns

Does your pattern work?

Figure 2.12

They describe a moment with 4- and 5-year-olds where one boy's reasoning was visible (Figure 2.12):

> *Two boys in the reception class had succeeded in making a continuous pattern with pom-poms around a rectangular mirror. This involved a repeating unit of three white pom-poms, two yellow and one blue, resulting in a wavy border around the rectangle. When Sue asked them if they could make it fit better, one of the boys said, 'I might change the pattern' and then removed one of the yellow pompoms from each unit of repeat.*

> (Thouless et al. 2020: 30)

This is an impressive example of a young child being able not only to recognise the unit of repeat within a pattern, but to generalise sufficiently well to adjust the entire pattern so that it fits around a rectangular shape – so that it 'works'. Take a moment to try creating a rectangular border pattern yourself to get a sense of the challenge as you navigate the corners and keep the pattern intact.

The adults in this classroom begun each morning with a pattern containing an error for the children to spot, describe and correct. Gradually, children would be asked to explain and justify their solutions. This project extended research into young children's awareness of mathematical pattern and structure by Mulligan and her colleagues (2008). Over 7 years of their research, the authors found that older children who were identified as lower-achieving mathematically *had a very poor grasp of mathematical pattern and structure* (Mulligan et al. 2008: 11). However, their subsequent research in kindergarten classrooms after pattern-teaching interventions revealed that young children could be taught to recognise and reason about pattern and structure, which consequently improved their overall mathematical learning. This can be done, as demonstrated by Thouless and her colleagues and the practitioners they worked with, in a playful, engaging way, by surrounding children with patterns to discuss and loose parts with which to pattern-make freely, discussing these and sensitively sandwiching these free exploration periods with carefully tuned moments of teaching.

Playful error-making can be used sensitively in many contexts to build children's reasoning skills. We can try:

- **Building** – placing two bricks of unequal height to build a bridge. *A child might*: swap your misplaced brick with one to match the other in height.

Figure 2.13

- **Completing jigsaws** – deliberately offering a child a piece that will not fit or offering one likely and one unlikely piece for them to choose one. *A child might*: ignore or refuse your offered piece or laugh and look for another more suitable piece.

- **Tidying up** – ostentatiously putting the incorrect number of pens into pots labelled with a specific amount, pegging up a set of 0–10 numerals incorrectly or storing the wrong block in a template on a shelf. *A child might:* point this out: *this is in the wrong place/shouldn't be here, there's the wrong number (in) here*, or physically correct the error.

Figure 2.14

- **Pattern-making** – deliberately changing one piece of a symmetrical pattern so that it becomes asymmetric, asking a child to *hide their eyes* whilst you make one change to their pattern for them to spot. *A child might*: point this out as: *not fitting* or *not looking right,* or physically correct the alteration.

Figure 2.15

Children in an error-spotting environment become quick to spot, correct and, gradually, explain their reasoning. We can ponder with them: *I wonder why … you think Serifa is saying that?* Gradually you can ask for a more *general* explanation, by saying: *Can anyone explain what Milo did wrong just there? Can you tell me what to do to put that right?* and once again, *I wonder why …?* We might not receive a clear explanation, certainly not immediately, but by asking the question we are beginning the process of drawing young children's attention to the 'why' and occasionally we can start to model the resulting 'because'.

GAMES

To encourage all young children to reason, 'low-threshold' starting points (open to every one of our children) are a way of ensuring every child is initially able to access and has time to fully engage. A necessary step is then to provide time for sensitive discussion with an adult to encourage children to reflect upon what they (and others) have done. A familiar game can be used in this way to engage children in some reasoning, as in the following example with some 5- and 6-year-old children.

VIGNETTE: PREDICTING IN A COLLECTING GAME

Five 5- and 6-year-olds are sat around a container of small sorting toys. They each have a dish and are rolling a die with +1, +1, +2, +2, −1, −2, on the faces. They are mid-game and each child has a different number of toys in their dish.

Figure 2.16

Boy 1:   I got a boat!

Emily: (*with six toys in her dish, rolls the die marked '+2'*) Add two.

Amy: Emily's gonna have eight.

Emily takes two toys from the central store and puts them with the six in her dish.

Amy: (*rolls the die, gets '−2', re-rolls and gets '+1'*) Add one, add one (*adds one and counts her dish of toys*), one, two, three, four, five, six, seven, I got seven.

Emily counts hers, Amy watches.

Emily: Eight!

Amy: (*enthusiastically*) Emily's nearly got ten! She's got eight and I've got seven.

A boy rolls the die and adds two to his dish of five.

Girl: Amy, it's my turn.

Amy: In a minute, Emily is going to win.

Girl: Add two, add two, look (*adds two to her dish*).

Boy rolls '−2', laughs, takes two from his dish of six and replaces them in the central container.

Games have to be played many times for the thinking to move on from the management of the game to being about the mathematics of the game. The possible mathematics of this game as it is played in the vignette above can be listed as follows:

- *Counting*: knowing the number names, tagging each item with one number word, subitising, (instant recognition of a small amount) recognising numerals and attaching to a quantity.
- *Cardinality*: knowing the last number said gives the total amount so far, conserving quantity.
- *Comparison*: identifying the same quantities, identifying more and fewer, reasoning about amounts.
- *Recognising and operating with addition and subtraction symbols*: on this occasion the die was marked with symbols as operations
- *Predicting*: what will happen if I roll ...?
- *Reasoning*: understanding the 'because' as well as 'how' an amount changes, e.g. *I will be first to ten if I roll a three because I already have seven.*

The simpler the game, the fewer the rules and the more familiar the game, the more opportunity there is to talk about what is happening and why; the *if ... then ... because.* The collecting game in the above vignette is very familiar to these children; they are able to play it for an extensive period of time, without an adult, in quite a large group. It is clearly enjoyable and, as it is familiar, Amy can be seen analysing the situation and making predictions about the amount in Emily's dish. She says Emily will have eight as soon as she sees her die roll, that she has *nearly got ten*, and predicts she will win. We can model the use of *because* when we have discussions about the game after the event, for example: *Why did you say that Emily would win?* and, *What would have happened if Emily had rolled … next time? Why do you say that?* Adults can model predictive reasoning whilst playing alongside children by making observations like: *If I roll a 'take two' next, I'm going to have none left! I'm going to have more than you if I roll a three;* as well as asking questions such as: *What would you* like *to roll next? I wonder why?*

Figure 2.17

Griffiths and her colleagues suggest setting up two toys to play a collecting game, with children rolling a die for each toy (Griffiths et al. 2016), as children can be more dispassionate with toys than when personally involved in playing the game against someone else. Moreover, it is easy to compare on a smaller scale how many each toy has and make comparisons between the two quantities. They can roll the die for each toy and be encouraged to chat about what happens. If we share an experience of playing a popular game, our maths chats at the end of the session can include prompts to children to describe, explain, predict and reason based on what happened that day: *I wonder what made you think that? I wonder what number Bear was wishing for next? What might have happened if Bear hadn't rolled a three?*

Figure 2.18

Whichever games we use, we *all* benefit from them being played repeatedly. As a practitioner, playing a game or introducing a favourite task over and over means we get better at spotting useful responses and at anticipating the opportunities to extend children's thinking after the event. We are more aware of what is typical or likely or interesting and this means I can introduce an idea on the basis of a raft of knowledge of previous observations of children playing that game, as well as my knowledge of where the children are in their development. This is what Shulman calls *pedagogical subject knowledge* (Cochran et al. 1993; Shulman 1986) – a blend of subject knowledge and pedagogy. For example, I know the 'Pass it on' game (Chapter 3, page 57) will lead to someone having an empty plate if I set it up so that we each begin with just a few objects. This will open up the opportunity for the players to reason through what the consequences might be of what they decide to do (you are 'out', you roll again). Alternatively, if we play with a die with a blank side (or that includes a zero) this will lead to a discussion about the consequences of what a zero might mean in this context (miss a go, pass nothing, roll again).

These discussions can take place with my awareness of the possibilities of the game and without me being there to observe the game every time.

STORY AND NARRATIVE PROBLEMS – COLLECTIVE REASONING AND GENERALISING

These are contexts that provide many opportunities for children to talk together and invoke reasoning co-operatively, through taking part in a shared narrative. The following example began with the Reception teacher observing many children in her class racing each other around a circular track. The teacher observed the play over several days and developed this by providing a lot of connected experiences, such as introducing digital stopwatches for them to explore and setting up and modelling the idea of a scoreboard. The latter resulted in regular whole-group discussions about what the children scored and led to more children becoming involved. The class was shown a video of Usain Bolt breaking the world 100-metre record in 9.78 seconds and the adult led a discussion about this being *nearly but not quite ten*. The children were inspired to time each other as they raced around their outside track. This they did paying attention to accuracy (when to start and stop the watch) by racing one at a time and calling a runner back if their start did not coincide with the stopwatch start. The adult interspersed her observations of this play with regular 'review discussions' with the whole class about what was happening, giving her the opportunity to assess their understandings and introduce new questions and ideas. After a few days, two regular participants, Mark (5 years) and Elliot (4 years 11 months) were invited to watch an excerpt from a video taken of their play and invited to explain what was happening: (see vignette 'Racing' on facing page).

This example is complex mathematics that 5-year-olds would not be expected to understand; that a larger number on the stopwatch indicates a slower time, and that it is the smaller number that is required. Elliot clearly enjoyed the inverse relationship between moving very quickly and approaching zero, testing his idea to its limit by saying: *If you went* really *fast you will get zero!* Something mathematicians – as well as young children in their play – often do. Both Mark and Elliot are able to reason about what they have been involved in by being given the opportunity to reflect on this. Mark's comment, *How would you get zero? You're not running, if you get zero!* is a clear example of reasoning. What is critical in nurturing this level of thinking is providing time for children to think about what they are saying. Instead of correcting Mark's first statement, *'if you get faster you'll get a bigger number'*, I carefully repeated his statement, paying attention to my emphasis and tone of voice to give no clue that this was incorrect.

This is an extraordinary piece of collective reasoning from two very young children. It has not come about by chance. This is enjoyable, self-chosen and self-directed play interspersed with discussions with an adult who has some possible structure in mind, with adult and children collaboratively developing this over a few days. Collaborative reasoning opportunities such as these come from enjoyable contexts with a developing narrative that makes sense, alongside skilful interaction from adults who have a sense of 'direction of travel' in terms of the mathematical development. It is 'low-threshold' in that exploring and comparing numbers within ten are perfectly pitched for 4–5-year-olds and, although the context was at quite a high level of abstraction, it was meaningful and developed over a number of days and thus became accessible for them. For example, the use of a scoreboard generated a body of data over these days and children could begin to spot a pattern between speed (winning) and size of number (small).

VIGNETTE: RACING

Figure 2.19

Adult: So, can you remember what you were thinking about with your scores?

Mark: Err, we was, if you get faster you'll get a bigger number.

Adult: If you get faster you'll get a bigger number? Can you tell me a bit more about that?

Mark: No – If you get *slower* you'll get a bigger number. If you go *faster* you'll get a *smaller* number.

Adult: Ahhh!

Elliot: And if you go *really* fast you will get a zero! (*waves arms in circular movement*)

Mark: No, you won't Elliot! (*They both smile*)

Adult: If you went so fast as …

Elliot: Pop!

Adult: So teach me about that, Elliot.

Elliot: If you went *really* fast you will get zero! (*'writes' a zero in the air*)

Adult: Help us learn about that; how would that work?

Elliot: If you move your arms go really fast, umm …

Mark: No.

Elliot: … you will get zero (*writes a big zero in the air with his finger*)

Mark: No! How would you get zero? You're not running, if you get zero!

(*Both boys laugh*)

(Continued)

Figure 2.20

Children were:

- given the space and time over days to explore an intriguing concept; in this case, 'who is faster?'

And

- regularly given the time to reflect both upon their play and upon the speeds generated by discussing and analysing the scoreboards each day and being encouraged to explain what they noticed.

Figure 2.21

Using video to reflect with young children in the way described above is something we would use periodically rather than daily, but it need not be time consuming, it can take place with a larger group and is a way of encouraging reflection and depth of thinking which is more available to us with smart technology. Viewing a clip of videoed play will vividly bring back the moment of involvement in the play for the children and I have found phrases such as *help us learn about that* and *Teach me about that* encourage children to say more about what they were doing and to explain their thinking. These work in part because they put the child in the position of expert.

Imaginative contexts can provide safe environments for problem solving and reasoning, right up into adolescence, as typified in the work of Twomey Fosnot and Dolk (2002) and the Dutch Realistic Mathematics Education (RME) project at secondary school level (Hough and Gough 2007). Both take common experiences such as purchasing or sharing items and pose these as imaginative conundrums as starting points for mathematical exploration, experimentation, questioning and reasoning. Nunes et al. (2009) maintain that, from the earliest mathematical situations children face, they must learn to do several things:

- pay attention to information in whatever form it is given to us (for example, in a story situation);
- pick out and remember the relevant parts of this information;
- remember facts and procedures that help us answer the question we want to answer.

(Nunes et al. 2009: 13)

One example from Fosnot and Dolk (2001) uses the context of passengers getting on and off a bus to model addition, and later still, subtraction. In this classroom, a discussion about bus journeys with 5- and 6-year-olds moved into free play at buses, passengers and drivers using

Figure 2.22

Figure 2.23

Figure 2.24

Figure 2.25

rows of chairs, a driver and children getting on and off. At an appropriate point, the driver was asked if they knew how many people were on their bus. The children had not been paying attention to this, so they re-start, with people waiting at bus stops, and the adult drawing attention to how many were waiting at each stop and children calculating the eventual total at the end of the route. This happened many times with the children increasingly taking control. The bus route was later represented by the children as a drawn line with them recording the amount of people getting on each time. After this activity some children chose to make their own money and to use it in their play on the bus whilst others chose to draw pictures of the bus and how many people were on it. Gradually, this particular imaginative situation is mathematised, symbolised and generalised, with children encouraged to predict *how many would be on board* if … this or that happened at the next stop, and to explain why.

FRAMING TASKS TO SUPPORT REASONING

Interacting sensitively whilst providing broad, mathematical experiences is key to developing children's reasoning. When considering exactly what broad experiences to provide, we also need to consider how we frame these experiences in order for us to develop our young children's reasoning. Here the Characteristics of Effective Teaching and Learning, statutory in the English Early Years Foundation Stage (Department for Education 2021), play a key role, alongside the development of children's *self-regulation* and *executive functioning*. Let's consider these separately.

Self-regulation. This is the ability to recognise and regulate what you are doing and feeling. It involves an awareness of our own mental processes and is crucial for learning (Whitebread and Coltman 2017). Moylett (2018) has distinguished two aspects of self-regulation:

- *Emotional* – our well-being as learners, how we feel about ourselves as learners;
- *Cognitive* – awareness and control over our own learning.

Both of these aspects are significant in the learning of mathematics.

Executive functioning. This is the set of mental processes responsible for how we control our own behaviour and work towards goals – how we filter out distractions, set goals and work to achieve these and prioritise tasks to do so. Executive functioning develops throughout childhood into adolescence and has three main elements: *inhibitory control, cognitive flexibility* and *working memory*. These relate closely to the three factors listed earlier (shown in bold below) as necessary when children face mathematical situations (Nunes et al. 2009):

- *pay attention to information in whatever form it is given*: **inhibiting** a desire to act impulsively and pay attention to possibly interesting but irrelevant information;
- *pick out and remember the relevant parts of this information*: **shifting attention** and selecting information that is pertinent;
- *remember facts and procedures that help us answer the question we want to answer*: **keeping the goal in mind** and monitoring progress towards this.

There are three statutory Characteristics of Effective Teaching and Learning (Department for Education 2021):

- Playing and exploring – children investigate and experience things, and 'have a go';
- Active learning – children concentrate and keep on trying if they encounter difficulties and enjoy achievements;

- Creating and thinking critically – children have and develop their own ideas, make links between ideas, and develop strategies for doing things.

(Department for Education 2021: 16, para 1.15)

Playing and exploring. This is about developing the skill to become engaged and absorbed, including in something mathematical. To become absorbed in some mathematics we have to be offered both opportunities to play with and extended time to become engrossed in measuring, building, pattern-making and counting, playing with our ideas and the ideas of others. We have to build confidence in ourselves as mathematical beings.

Figure 2.26

Active learning. This is about having the resilience to keep going. Here our *executive function-ing* plays an important role. Without being engaged or being resilient enough to overcome setbacks, children will not experience the thrill of developing their own mathematical ideas.

Creating and thinking critically. This is the thrill of discovery and of children pursuing their own ideas; and is of particular importance to the development of mathematical reasoning. It rests on the other two characteristics.

With a focus on these three characteristics in our mathematics provision, reasoning can flourish. Children need to be given opportunities to make decisions about how to go about something and time to develop the facility to lose themselves in their activity. They need to be encouraged to ask questions as well as answer them. They have to learn to persevere and to consider alternative strategies and be supported to anticipate, visualise and form predictions to test out. They need to be helped to review and eval-uate their progress.

Here are some examples of simple provocations, originating from observing children playing, that integrate the three characteristics of *explore, persevere* and *think critically.*

- **Can this small pot hold 100 of anything?**

Collections of loose parts and a variety of containers to fit these into is a naturally engaging context in which to explore some mathematical reasoning. With younger learners, providing identical small containers and a range of 'treasure' to fit inside is one starting point. Leave these out with sticky-notes and pens and listen to and observe what plays out. Gradually, we can ask questions like: *How many things do you think you can fit inside? What are you going to choose and why? Which has the most buttons inside, I wonder? I wonder if we can fit exactly five … ten … 20 … in here. Why do you think that? How could it hold 100? Why do you think that? What would you like to try next? Let's see if you are right tomorrow. I wonder what might happen if we used these smaller beads?*

Gradually children can be encouraged to collaborate with a partner to choose bags or boxes containing differing quantities, and to work together to plan how to count them and then record their count (Gripton and Pawluch 2021). Having containers with divisions such as egg boxes or ten-frames to support their counting are useful. *How did you decide to count these? What did you decide to do next? What did you find out? Tell me more about your recording. What might you do differently tomorrow? Why is that?*

Figure 2.27

- **Can we make this sheet of card roll along the floor?**

Making something as angular as a sheet of paper or card into something curved is counterintuitive. Model rolling up an A4 sheet to make a cylinder; this is tricky and will need some practice as a cylinder has sides that are parallel. Rolling into a more cone-like shape will lead to something different happening and different questions. *Does it roll? Is this a cylinder? Why? Why not? How could you make it roll further, do you think? What could we try?*

> *Without using scissors: How might we make a cylinder 'family' of many different sizes? How might we organise them?*

Figure 2.28

- **Is this a triangle if its three sides are all different lengths?**

Explore making two-dimensional shapes laying down sticks of different lengths. *What shapes can you make? What do you notice about them? How might we sort them out into different types? Where are the squares? Which are triangles? What is the same, and what is different about these two triangles? How do you know these are all triangles? What is special about a triangle?*

> *Try using Cuisenaire rods. I think there are some that cannot touch ends to make triangles; can we find three rods that don't 'work' as triangles?*

Supporting children's mathematics using the Characteristics of Effective Teaching and Learning is explored further in Chapter 4 ('Adults – What do we do?').

Figure 2.29

As executive functioning skills are key mental processes for supporting logical, purposeful thinking, this next example reflects on how we might encourage this in our work with young children.

Most children enjoy water play, pouring and filling containers, and do so with great enthusiasm. However, attempts by adults to 'mathematise' this play and discuss how we know when something is 'full', how much 'emptier' something is, or which of two 'holds more/less than …' may not gain their interest. Often children seem to be paying attention to the container that is being *emptied*, rather than the one being filled. To move the play on mathematically, we might make several changes:

- Ensure there is a range of transparent containers that a child can lift in one hand.
- Provide a table for the filled containers to sit on.
- Colour the water in order that the contents can be easily seen.
- Mark each container with a masking-tape line (this is to be the 'fill up to' line).
- Provide jugs for pouring.
- Work with the children on a context for the water play – in this example, a lemonade factory, following the reading of a picture book.

VIGNETTE: A LEMONADE FACTORY

Figure 2.30

(Continued)

The class is read *Little Croc's Purse* by Lizzie Finlay, a story of a lost purse and lemonade. After enjoying the story together more than once, play develops around the water trays involving 'lemonade'. It is thirsty work and soon there is a 'lemonade shop', with other children entering the play to 'drink' the lemonade, asking for their bottle to be re-filled. The adult plays alongside the children, demonstrating careful pouring from the jug, up to different 'fill up to' lines. Children play with this idea; they are encouraged to call out when the 'up to' line is reached. Discussion ensues about different containers and how much lemonade they are holding. Children practise careful lemonade pouring, taking it in turns in pairs, one holding their chosen container and the other with a jug of lemonade to pour. The child holding the container watches it fill closely, in order to call out *stop!* when the line is reached. The bottles of lemonade are lined up on the table. Discussion takes place about whether the containers are filled 'fairly', not full enough, fuller than others, 'nearly full' or too full to carry (or 'drink'). Later, children decide which prices are charged for different amounts of lemonade.

Figure 2.31

How is this play encouraging children's executive functioning skills?

Firstly, ***inhibitory control***: the addition of a 'fill up to' line means children have to stop themselves (and each other) from over-pouring, or anything else they might want to do with the water on this occasion. They have to pay attention to what is relevant by focusing on the container being filled rather than the one being emptied, and to the point they have to stop pouring. By taking on the different roles of receiver and pourer, they are inhibiting any preference they may have for one of these, or to not co-operate at all. The introduction of a purpose for the pouring means children have to act in a way that is sensible in terms of that context.

Attention shifting and cognitive flexibility are exercised by the shifting of roles from pourer to receiver, as well as the fact that children must shift attention between the container being emptied and the container being filled and that different containers have 'fill up to' lines in different positions. As the masking tape has width, they have to decide whether to call *stop!* at the bottom or the top of the line, and how accurately they can do this. They have to be flexible interacting with the 'purchasers' of the lemonade in order to agree on the amount that is poured each time.

Figure 2.32

Children's **working memory** is being exercised in remembering what they have to do, planning what they do in the correct order, and recalling what happened and what they found out.

Mathematically, both the purpose for pouring and the 'fill up to' line focus children's attention on being accurate. The 'fill up to' line draws children's attention to how full the container is, and with this, increasing complexity in comparative, descriptive language can be introduced: *nearly as full as ...*, *just about half-way ...*, *too full to carry or pour ...*, etc. Selling the lemonade requires reasoning about what fairness might mean in this context; maybe we are going to need to pour the same amount in each container? How could we do that?

None of this is to undervalue full-on, open-ended, splashy water play, but does challenge the idea that water play ends here, and that mathematical comparisons of capacity occur naturally from this, which in my experience, they may not. These developments to the play both foster some deeper mathematics and integrate executive functioning skills. Joswick and her colleagues describe how we might consider both the mathematical demands of a task, and the executive function demands, referring to this as 'double impact': [the] *simultaneous development of young children's mathematical proficiencies and executive function skills* (Joswick et al. 2019: 418).

TO SUMMARISE

What is the biggest shape?

How thin is thin?

Is 100 the biggest number?

Questions like these have been asked many times by young enquiring minds observing and thinking about what they notice.

(Borthwick and Cross 2018)

Borthwick and Cross (2018) formulate a clear difference between *thinking* and *reasoning*, arguing that reasoning is logical, purposeful thinking. Young children are perfectly able to reason and to develop their mathematical reasoning, but they need to be presented with mathematical situations with simple parameters that make sense, and importantly, structured opportunities to reflect on these. Young children do naturally think about, reason and question the world around them – continually asking *why?* We need to channel this into their mathematics learning by providing interesting mathematical experiences to ask *why?* about. The answers to these questions are not as important as opportunities to engage in asking questions and to practise reasoning about answers. We can then monitor children's application and focus – are they getting nearer their goal? It is often the atypical or the surprising that triggers reasoning in an attempt to understand it. We can use the question *what do you notice?* to stimulate initial thoughts and follow this with *what do you wonder?* to encourage exploration.

Figure 2.33   What do you notice?

Reasoning depends upon us not ending the mathematics with the task itself, but on the adult making time with children to discuss what happened and what was noticed. Mathematical proof suggests a formal, rigorous procedure, but posing questions such as: *Can you prove that one is longer, … heaviest, … that everyone has the same amount?* is important in order that children become comfortable with the idea of justifying their thoughts and findings. Mason et al. (1982) prefer to see proof as an attempt to *convince*, or to cause other people to *see what you see* and this makes the idea of justification and proof more accessible to younger learners.

Executive functioning and self-regulation are important in the development of children's mathematical reasoning. We can design tasks so that children *want* to inhibit distractions, and by building these on their own interests and activities, it is more likely they will be engaged and persist. Opportunities to re-visit tasks, for example, playing a game over and over in order to draw attention to patterns that occur, help children move beyond the activity and reflect on what might happen. Carefully designing mathematics tasks to cause young children to think flexibly in order to solve a problem and then encouraging them to analyse the results are essential mathematical experiences for promoting mathematical reasoning. The following chapter ('Preparation rather than planning') considers how we might prepare for these experiences.

Characteristics of mathematics are evident in young children's activities. If we teach mathematics only in the narrow senses of passing it on passively as a body of knowledge, or formally as a symbol-system, we are doing a disservice both to the nature of this human activity and to young children's natural capabilities and considerable powers.

(Bird 1991: 171)

3

PREPARATION RATHER THAN PLANNING

'You know, I like school, but there's no time to do anything.'

WHAT DO I MEAN BY PREPARATION?

This 4-year-old has perfectly articulated what it feels like to have your day filled with other people's learning, which she sees as nothing to do with her interests or enthusiasms. How can the experience of mathematics become something that the child feels connected to and involved in? That is the theme of this chapter.

I am a teacher and therefore I teach. I introduce things, I suggest, I instruct, I tell. I also observe, respond, change tack and listen. I probably do not remain silent often or long enough. The forging of links between the curriculum and the child is an organic process. This resonates with what Marion Bird refers to as planning for *foreseen possibilities* (1991: 133), where an adult anticipates what *might* happen when planning an activity but does not let that restrict what does happen in practice. This is what I refer to as *preparation*. Historically, planning, as opposed to preparation, leads me to think through, and usually commit to 'paper', my teaching sessions, from beginning to end, my provision and environment, some of what I will do and what I will say; what I will provide, probably with a particular learning goal or intention in mind. Preparation, on the other hand, is about having researched the area I am teaching well enough to be sufficiently confident to go off in different directions from my starting point, when the moment demands this, and secure that I will 'cover' the mathematics that needs to be covered over time. 'Over time' is critical here, in a phrase attributed to mathematics educator John Mason: *Teaching takes place in time; learning takes place over time* (Griffin 1989). Preparation includes spaces for the children to follow their interests from my original stimulus, and time with my minimal interference for them to make sense of what they are being taught. Preparation involves me being flexible about what I plan to say, do and provide, re-planning as we go along together. Preparation allows for learning to take place over time rather than in shorter 'units' of time. Marion Bird puts this well:

> When I am actually working with the children and they seem to be purposefully engaged, I try and focus away from my earlier thoughts on how the situation might develop and concentrate on what the children actually seem to be doing and suggesting.

(Bird 1991: 133)

Figure 3.1   Counting into a container with sections.

Often what the children actually do provides me with new ideas and directions for us to explore together. Bird goes on to say: *However, my earlier ideas can be brought into the open if, for example, providing a new focus to the session feels appropriate.* Allowing for space and fluidity is at the heart of being playful around mathematics, stopping to have a play with, explore, something that becomes interesting to us. Over time, we can choose to focus on different mathematical objectives by making adjustments to what we provide and in how we choose to focus our discussions with the children (Serret and Gripton 2020). For example, consider the collecting game vignette ('Predicting in a collecting game') in Chapter 2 (page 34). The adult here has introduced the focus for these 5- and 6-year-old children by providing a particular die, as well as the object of the game being the first to reach ten. Changing the die and the object of the game will alter the mathematics I wish the children to experience.

For example, with much younger children our main objective might be recognising, counting and comparing small quantities accurately. For this, I will use a die dotted to four (to encourage subitising), play the game with a small number of similar items (e.g. coins or cubes in one colour to avoid distractions of colour and type) and perhaps give each

Figure 3.2   Choosing loose parts and containers for playing a collecting game.

child a container separated into sections (e.g. an egg box) in which to place each item, to aid recognition of small amounts and the relationships between them. There are many small adaptations that can be made to this simple game over time that draw children's attention to different mathematical ideas and to deepen their experiences in counting and comparing amounts. Thus, I can shift the emphasis onto different mathematical objectives, whilst emphasising the links between these by keeping some things constant.

Observation is a key aspect of being prepared. By observing how the children's mathematics develops, both independently and with adult support, I can build more effectively on what my children know and can do, moving forwards. Another key element is to deliberately prepare to be playful. Both of these are explored in more depth in the examples that follow.

PREPARING MATHEMATICALLY FOR ROLE PLAY

Figure 3.3

Figure 3.4

Most Early Years settings have a 'role play area'. Role or pretend play can be defined as a child temporarily 'walking in another's shoes' (Williams 2006); it has a recognised and significant social component: *the shared pretend play between children in which they temporarily act out the part of someone else using pretend actions and utterances* (Harris, in Rogers and Evans 2007: 165). However role play, even if we plan for this to include some mathematics, often does not appear to engage the children in thinking and working mathematically. Whilst Gifford (2006) points out that some scenarios, such as birthday party role play, can be mathematically productive, in the opening sentence of her book she observes that:

> *I used to spend large amounts of time and effort creating shops with five-year-olds so that they could learn mathematics through play. ... whenever I looked, they were stuffing bags full of goods without paying (i.e. shoplifting) or paying arbitrary amounts and then demanding huge quantities of change from the shopkeeper.*

> (Gifford 2006: 1)

Figure 3.5

After reading this, I became involved for some years in investigating how role play might become more effective in enriching young children's learning of mathematics. It seemed to me that narrative, pretend play and storifying were important aspects of Early Years teaching and learning: why should mathematics be excluded from this powerful learning context? Play is a safe space for children to try out their emerging ideas, and being fantasy, mistakes are not of great consequence. Learners thrive if they have a safe space to try out ideas and time to make sense and to practise what they are learning. Alongside opportunities to make and correct mistakes, these are essential for the development of healthy mathematicians with a confident approach to new mathematical situations (Gifford 2015).

In the sections within this chapter on block play (page 70) and number play (page 91), I outline how narrative can provide a meaningful context for playing with mathematical ideas. Role play is the most immersive of narrative play. The example that follows illustrates how role play might provide an effective context for learning mathematics.

VIGNETTE: THE DINOSAUR CAFE

Figure 3.6

Luke and Edie are 'dinosaur customers' in a cafe set up for dinosaurs. Luke is holding two sticks. Cathy and Lucy are serving them. All the participants have freely chosen to play here. They are 4 and 5 years of age.

Lucy: Are you full up now?

Luke shakes his head (no). Luke strikes the two sticks together firmly nine times.

(Continued)

Lucy: Is that nine?

Luke nods and Lucy turns to fetch him something.

Lucy: It has to be one of these (*laughs and hands plastic chicken to Luke*).

Luke 'eats' the chicken.

Lucy: Are you full up? (*laughs*)

Luke shakes his head and strikes the two sticks with care, ten times. Lucy can be seen counting alongside.

Edie: He'll eat all the food in a minute!

Lucy: Ten?

Luke nods. Cathy offers him a pineapple; Luke doesn't take it.

Lucy: (*consults the price board, laughing*) Sweetcorn.

Lucy hands Luke a corn on the cob. Luke 'gobbles' the corn.

Edie: Are you full up now?

Luke shakes his head and Lucy laughs.

Figure 3.7

This extract was part of a much longer episode. We can sum up the mathematics here as follows. Customers decide what they want and refer to a menu board, where food items (decided by the children) are numbered one to ten (an idea introduced by the adult). Dinosaur customers read the numeral and because dinosaurs cannot speak, communicate their choice by tapping together two sticks the appropriate number of times. Those working in the cafe count the number of taps the customer makes, link this amount with a written numeral and then to an item of food. In both roles children are challenged to count something that cannot be seen, i.e. sounds.

How did this come about? The practitioner had observed the children's play around exploring and catching wild animals and the practitioner responded to this by providing equipment (camouflage, binoculars, factual information on the identification of animals, etc.) and the children had wanted a cafe for the explorers to visit. One day an explorer spotted some dinosaurs; a 'dinosaur' went into the cafe and created a furore. As the explorers were not happy with this, a child's idea of a cafe serving only dinosaurs grew from this play. Although ordering, serving, eating (and sometimes paying) for food are fairly typical in role play, this is an interesting example because the mathematics involved is not to do with exchange by payment (in fact, no payment took place at all as '*dinosaurs don't have money*'), but instead the mathematics was about accurate, numerical communication. In discussion with the children the adult has genuinely asked how a dinosaur could possibly order food and the idea of banging something emerged. She used this as an opportunity for some counting practice, and in this context, it makes perfect sense. An important element of this is the adult's willingness to go along with the children's play, however strange it might seem. She describes this as:

> [We are all] ... in a complete imaginary world ... all completely talking at the same level.
> ... When you introduce something mathematically, then they don't question that either.

(GH, in conversation 2012)

This occasion is mathematically successful because the adult is alert to opportunities to sensitively introduce some mathematics at a pitch the children are working at, and it builds on what the adult has observed as well as what the children suggest. Regular class and group discussions feed into the play, developing it further. This type of role play is inexpensive and does not require time-consuming resourcing: for example, on this occasion, there were no dinosaur suits provided; the children simply roared.

Play that incorporates plot and story line is a reflection of a child's developing notions of the world (Garvey 1977). This is play flowing into literature, an approach encapsulated in the experiences described as 'helicopter storying' (Paley 2004). Paley's view is summarised in the following sentence: *fantasy play is the glue that binds together all other pursuits* (Paley 2004: 8). In the dinosaur example above, we have play flowing into mathematics, and vice versa. For this to happen, we have to be alert – and prepared – for mathematical possibilities and then structure these carefully and sensitively with the children.

Taking the common example of shop play, Griffiths (2010) gives sound advice on making this mathematically more effective:

- Confront issues of economic value.
- Provide sufficient time for children to develop their ideas.
- Discuss the children's work with them.

We will look at these one at a time.

ECONOMIC VALUE

Figure 3.8

A desire to 'make the arithmetic manageable' sometimes leads to teachers restricting prices to small amounts in pennies, cutting across children's growing experience of what things cost. A better alternative is to allow children to choose their own prices for goods in their shop. This gives them the opportunity to work at a level they are comfortable with; they are often more ambitious and more successful than their teacher expects.

(Griffiths 2010: 22)

Having a charity shop or a car-boot sale is an authentic way to keep prices down, as does opening the ubiquitous 'pound shop', which has the additional advantage of dealing with whole numbers. Having access to an endless supply of plastic money is unhelpful if we want to focus on accurately counting coins as it becomes valueless. I have found it effective to give each child, or pair of children, responsibility for a purse or wallet containing a set amount of money that has to last for the day (or gradually, a few days or a week). Not every purse contains the same amount; after all, when do the contents of any two wallets exactly match? I have found this encourages children to pay careful attention to how much they spend, what they choose to spend it on, how much they have left to last the week, as well as with older children, how much change they receive. These constraints focus attention on ideas of cost and value and children above the ages of 5 and 6 can benefit from being involved in carefully crafted shop role play such as this. In the vignette that follows, the economics are real and a deal is struck.

VIGNETTE: THE CHARITY SHOP

Figure 3.9

(Continued)

Figure 3.10

Shop volunteer
(aged 6 and on
the phone):   You'll never guess, this person wants to come and drop things
                          off at 1 am! Who do they think I am? They'll have to come
                          later.

Shopper (aged 5):   I want this.

She wants to buy a treasure chest for £9 but only has £6 as she has already bought
two flowers for £1 each.

Shop volunteer: Hmmm, let's see if we can do you a deal. Give me the flowers
                          back and then I'll give you the treasure chest for £6.

Figure 3.11

ALLOWING TIME

Role play does not necessarily mean you need to set up special areas in your room – older children can work at their usual desks or tables, taking turns at being the customers or staff of a post office, toy shop, travel agent, car show room or cafe.

(Griffiths 2010: 20)

In terms of time spent role playing, for older children there can be a shift from the extended, uninterrupted periods necessary for younger children and in settings where time is more fluid (Rogers and Evans 2007) to shorter periods spent over the duration of a couple of weeks, or daily 'shopping time' at tables. Older children have been found to particularly value the time and the privacy role play affords them to explore the mathematics they are being taught (Ross 2011). Flexible use of space, including the use of outdoors, can aid such privacy. We also need to be prepared to allow time for Griffiths' third point, group discussions about the mathematics and to build on participants' ideas.

DISCUSSING THE CHILDREN'S WORK WITH THEM

With all ages, listening to children's views about play preferences, along with periods of observation and joint review, raises its value (Rogers and Evans 2007; Williams 2014). Time 'out of role' and away from the pretend play, where children and adults extend their mathematical thinking together, is critical for such play to become effective in developing children's mathematical thinking. This is as true for younger children as it is for older ones. Such time gives status to the play and stops it from becoming stale, a space where both children and adults introduce new directions and ideas. The 'Mantle of the Expert' work of Dorothy Heathcote, her process drama approach with older children across all curriculum areas, which makes use of fantasy contexts to tackle problems (Taylor 2016; Wagner 1999), similarly advocates moving in and out of role to develop both the story and the learning. Discussing what has happened together provides fresh problems to solve. Children can be encouraged to take an active role in developing the mathematical ideas as well as in the play itself.

PREPARING FOR BLOCK AND SHAPE PLAY

Play with unit blocks related in size and shape integrates all the key spatial aspects we need to develop in children, for example:

* understanding relationships between shapes, sizes and part–whole relations;
* developing visualisation and memory;
* extending language;
* understanding position and orientation;
* understanding transformation and rotation.

Large blocks are a key component of many Early Years settings and are an important way to develop early spatial and mathematical thinking (Gura 1992). Those of us lucky enough

Figure 3.12

have a complete set of solid wooden unit blocks from Community Playthings: **www. communityplaythings.co.uk/products/block-play**

Unit blocks all have a relationship to each other, so that, for example, two of the cubes will match in volume the rectangular prism and two of the 'wedges' will fit together to match the volume of rectangular prism, and so on.

Pattern blocks are also based on a unit system and are a rich resource for exploring mathematical pattern and are discussed a little later in this chapter.

Developing young children's spatial reasoning supports their number learning. Spatial and numerical reasoning influence each other, they are intertwined, and research suggests that spatial reasoning is indicative of much later, overall mathematical achievement (Cheng and Mix 2014). Research is now overwhelming regarding the positive relationship between spatial reasoning skills and wider mathematical skills, with this relationship emerging from as young as 3 years of age (Schmitt et al. 2018; Verdine et al. 2017). Time spent developing

Figure 3.13

spatial thinking is a sound educational investment. However, block play can still be periph-
eral to the mathematics curriculum and lack development. Moreover, gender differences in
spatial play have been found to arise from an early age and have also been observed across
a wide range of spatial tasks, including: paper folding, puzzles, map reading and two-
dimensional mental rotation tasks, with boys generally doing better on these tasks than girls.
So, developing spatial confidence at an early age may be particularly critical for girls.
However, research suggests that gender differences in block play can be put down to the
amount of time boys spend playing with the blocks and are therefore more to do with *preference* rather than competence (Casey et al. 2008). Open-ended play with blocks is
extremely valuable, especially with an interested adult playing alongside to discuss and
extend children's language and thinking (Ribeiro et al. 2020). Examples of adult talk that are
particularly valuable refer to the properties of the blocks: *Look, here's another straight-edge piece.*

Figure 3.14

*This brick is the same shape as that one, isn't it? We could turn it around and see if it fits?
What size/shape do you need now?*

However, one particular piece of research throws up findings that have implications for
our block play with young children. Research into playful interventions with 5- and 6-year-
olds from Casey and her colleagues (2008) points to an interesting development. The team
found that introducing an oral story-telling context into the block play improved building
performance both in terms of the complexity of the structures and children's long-term
engagement in the play and that this was particularly the case amongst the girls.
The emphasis was on jointly creating a meaningful context for collaborative construction,

Figure 3.15   Block storage.

with the adult scaffolding the play that emerged. Tasks were centred around a character, Sneeze, who had, for example, been asked to build an enclosure around a castle with a wall to keep any animals from jumping over. As the story unwinds the children are asked to adjust the building to meet further conditions, such as building a bridge for the king and queen to cross the moat and travel into the castle. The key aspect of the story-telling context that seemed to make it so effective for improving block play was that the story provided both problems, and *reasons* that made perfect sense, for making the walls taller or the bridge wider. The story character set out the criteria for exactly how the children should build their structures according to specifications, which were consistent within the logic of the story and were an integral part of the storyline. Moreover, it provided a story character that children could sympathise with and wish to help, so supporting motivation.

There are parallels here with discussions, above, on role play (page 60), where, from a developing narrative, a problem emerges which has to be solved. Casey and her colleagues (2008) make the point that open-ended, unstructured block play favours males, whereas females benefit from a goal-directed scenario. This does not mean that open-ended block

Figure 3.16   Block storage.

play does not also have a crucial role to play, but, as was discussed earlier in relation to water play, it does indicate that it might not be enough and that we need to consider how and in what ways we develop free play. A number of studies have shown that using 'goal-oriented tasks' improves young children's overall mathematics, including their number understanding and general thinking skills (Verdine et al. 2017). Goal-oriented tasks, for example, building a bed for teddy or a car suitable for a superhero, can exist alongside free play and be developed from the children's own play and interests. These can then be judged by the children to see if they satisfy their aims.

Figure 3.17 Pattern blocks are related in size.

Figure 3.18

Figure 3.19   **Figure 3.20**   **Figure 3.21**

Shape and pattern play is a rich context for exploring spatial relationships. Pattern blocks are a particularly powerful resource and extremely attractive to children. Gregg describes them as *mathematically structured loose parts* (Gregg 2020: 3), as they have mathematical relationships built into the set. These, like the wooden building bricks discussed above, are unit blocks and as such they are a fairly unique pattern-making manipulative and can be the start of some playful and powerful mathematical thinking and reasoning. All of the images here were created by children.

As with any resource, children – of all ages – benefit from extended time, repeatedly, to explore the potential of the blocks:

> *The blocks are sufficiently mathematically structured that children's unstructured play*
> *would lead to all sorts of mathematical experience.*

> (Gregg 2020: 4)

Figure 3.22   Perfume bottle shapes.

Children playing freely with the blocks make a wide range of pictures and patterns, the potential of which Gregg says is limitless. For example: 'floors' and 'roads' can lead to tessellations and reflections, and making shapes 'bigger and bigger' leads to invitations to explore the nature and properties of two-dimensional shapes as well as number patterns. All of the patterns shown here have resulted from children's extended play with the blocks over time.

The key to enriching this play is, like block play, a combination of free exploration, with the adult observing what is intriguing the children to prepare to develop later, alongside goal-oriented play, such as: *What different perfume bottles/triangles/stars can you make? How are they different? How are they the same? What do you notice?*

Figure 3.23

Figure 3.24

PREPARING FOR WORK ON MEASURES

When we deal with numbers in everyday life it is often in a measurement context and children's early experiences of number often occur in measures contexts such as discussing distance, time and baking, for example when discussing how long it is until …, how far it is to the park, how much milk I am allowed to pour, how big a piece of cake I can have. An alternative view of the teaching of number, where number teaching is based on a sound foundation of measure and comparison experiences, stems from Davydov, a Russian psychologist. For Davydov, comparing in measures underpins comparisons of quantity. Davydov built on the ideas of Vygotsky (1978) and developed a curriculum where the teaching of measurement comes before abstract number. In Davydov's view number is first introduced in a measures context as a way of representing relationships, for example, this box *holds more* pebbles than this one. His argument is that this is natural and instinctive for young children because they make sense of their world by investigating such everyday relationships as: *I have more milk than her*, *This bag is very heavy* (Venenciano and Dougherty 2014).

Davydov's work remains largely unexplored as it is largely untranslated from the original Russian. The *Measure Up* elementary mathematics research project in Hawaii that began in 2001 continues this work by focusing on developing algebraic thinking through measure. Algebraic thinking includes recognising and analysing patterns, studying and representing equality and relationships, making generalisations and analysing how things change: **https://manoa.hawaii.edu/crdg/research-development/research-programs/mathematics/measure-up/**. More recently and following Davydov, Cheeseman, Benz and Pullen's research (2018, in Downton et al. 2020) describes an alternative curriculum for 5- and 6-year-olds, approaching all number through measurement-based tasks:

> *[this] allowed the children to explore number in meaningful ways and stimulated some children to go beyond the intended curriculum in relation to measurement outcomes (use direct and indirect comparison), to understanding iteration of units; and quantifying and comparing measures.*
>
> (Downton et al. 2020: 11)

'Iteration' refers to the use of a unit several times to measure something, as in counting the number of straws it takes to match the length of a tape. Their findings indicated that providing a measurement-focused approach led to an improvement in children's counting and place value knowledge, suggesting that approaching number through playful measurement contexts is valuable for young children.

Figure 3.25

Figure 3.26

Figure 3.27

LENGTH

In this example I have prepared some work on measuring length some work on measuring length by greeting the children as they arrive and handing them a length of tape or ribbon: *Good morning, Adil, how are you? Here is your measuring tape for today. We will be doing lots of measuring this week.*

How have I prepared for this? What do I anticipate happening? I have chosen tape and ribbon because of their resemblance to a tape measure. I have cut ribbons of

Figure 3.28

different (fairly random) lengths, to begin with that differ markedly in length. There are exactly two of every length in order that I can invite children to *find the one that is exactly the same length as yours*. All are long (short) enough to be stretched between a child's arms where they can still hold both ends, to make comparison straightforward. I have thought about (prepared for) what I think is important mathematically for us to discuss as the days pass:

- how to compare two lengths accurately and describe these;
- the necessary skill of matching 'ends' with care to make an accurate comparison;
- understanding equality and inequality;
- what 'big' or 'small' might mean, and opportunities to extend the comparative vocabulary of measures;
- how we might compare the different physical attributes of objects: length, height, width, depth …;

Figure 3.29

- identifying the need for precision;
- identifying *how much* larger/longer/wider is one thing than another.

I can identify a broad progression, for example, refining 'big' to 'long', long to 'a little (a lot) longer (shorter) than', maybe moving towards the more sophisticated comparison of the length of two objects *indirectly* using a tape or ribbon, i.e. *This log is longer than this table* because *my ribbon doesn't reach the end of the log*.

Figure 3.30

First, I need to listen to and observe what the children are saying and doing. Sometimes I see them meeting up to lay their tapes alongside one another or holding them up in the air to compare them. They spend a lot of time saying they have the *biggest*. I notice that they are not often matching these one at end to do this. I make a note to introduce ribbons that are close in length and thus have to be compared with more attention to accuracy. I observe one boy has tried to commandeer all the tapes to reach across the outside area. No one is as yet using their measuring tape to compare with a second object, such as a table leg or block model. We all meet on the carpet a little later in the day and I hear this:

I have the biggest.

No; I have a long one.

He took mine.

I can choose which of these children's statements to repeat and leave time for a child to respond. A discussion ensues. And within that discussion I can choose to introduce one or two of the ideas I think are important, such as: *How do we compare things accurately, so we are sure one is longest?* I deliberately hold my tape and a second one so that the ends do not match and state that mine is a lot longer. I await the indignation. I exaggerate the comparison if this does not happen. This is being playful and deliberately provocative. I move the tapes a little so that they *nearly* match. Someone has to instruct me how to compare two lengths accurately. We all have a practice. This might be enough for now. I might say: *Did anyone do any measuring and find something that was* exactly the same length *as their tape? How can I be sure they are* exactly *the same?* I might set that task for the next session. Meanwhile, I consider cutting more measuring tapes, all of the same length, to put in a 'measuring toolkit' outside, later including tape measures and rulers, in order that the play I observed, the placing of the tapes end to end to reach across the outside area, can continue and be developed. We could follow up this idea later to count how many tapes stretch from place to place, and to compare distances.

What is important here is that I am clear about the difference between *being prepared* and having well-prepared lessons (what I am referring to as 'planning'):

having well-prepared lessons: feeling in control and having a good idea of how the lesson will go; minimising the chance of something going wrong and being well prepared for a lesson: having good subject knowledge; knowing misconceptions; knowing your pupils and being open to situations that will challenge everyone in the classroom including you. Being well prepared for a lesson allows you to respond to feedback from your learners, without which you cannot have any sense of whether learning is likely to take place over time.

(Griffin 2019: 17)

In my length example I have prepared by making myself aware of how children's understanding of measure develops (my subject knowledge) and am thus clear about the experiences of measurement to which I am introducing the children and how I might intervene to move this forward. I prepare to move from direct comparison of one tape with another to indirect comparison with a second object. And later:

- experience of standard units and a range of measuring apparatus through play (for all measures), e.g. tape measures and rulers of all sorts, balance scales, weights, spring balances, timers etc.;
- exploring the length of one tape using a variety of similar units, such as the number of straws, cubes or pens that match its length;
- exploring different-length tapes using the same unit, e.g. how many flowers fit along this tape? What about this one?

Figure 3.31

The ribbon/tape scenario can lead into work with units quite naturally. Three key concepts at the foundation of mathematics are *equality, inequality* and *unit*. In this case, we can begin to explore *how many times* they think a shorter strip will fit into a longer one and by keeping the longer one (the 'container') constant, we begin to focus on the impact the size of a unit has on the count or measure. Playful and even ridiculous statements can be made, for example: *I think 1000 of my little straws will fit along your ribbon! Who can find the very most/least number (of things) to fit exactly along here? I bet my tape and your tape together will reach all the way from the door to that tree right over there.*

This example uses a measures context (in particular, length) for exploring comparison. It would be equally possible to replicate these ideas in a mass (weight) context (see below). Developing children's understanding of measurement is far more complex than a linear route from using non-standard units (e.g. Unifix cubes or stones) to standard (metric) units. At all stages we have to move 'backwards' to prior learning as well as forwards to future learning to help children make links and secure deep understanding. Most children have held or seen a tape measure and know measurement has something to do with numbers, but what do the numbers refer to and what are we counting? Having a box of authentic measuring instruments for children to explore helps me observe what they know, what they are able to do already and what I can build on. Maybe they can measure themselves using a height chart? Maybe they can make their own height chart? Can they use metric weights and balance scales with the playdough or in the mud kitchen to make pies?

In preparing to teach (as opposed to planning) we keep in mind that understanding is not linear and we will develop children's accumulating knowledge in different directions. I am clear of the core understandings I want these particular children to grasp and which are the experiences that are built on later – moving forwards and backwards across the mathematics.

MASS

Alongside lots of free exploration with measurement tools, such as the balances shown here in the mud kitchen, I have found home-made spring balances such as those pictured on page 91 effective in enriching children's understanding of comparisons of heavier than, heaviest, lighter than and lightest, because they allow us to compare more than two items. Children can enjoy filling these cartons, which are suspended on (responsive) elastic, with different items to see how 'far down' they can make them stretch, and thus find the heaviest item. To make one of these, cut a juice or milk carton in half, punch a hole in each side and thread this 'basket' with string. Then suspend this on thin elastic bands and tape them on to a wall or cupboard: **www.youtube.com/watch?v=YUDzLOVPZM8**

Provide a selection of small everyday items of different weights, for example, orange, potato, glue stick, small sock filled with sand, to place in the scales and compare. This can provide opportunities to recognise and analyse equality and relationships between the items, and some surprises. *Which item is heaviest (stretches the furthest)? Can we find two items that weigh (stretch) the same?* Having more than two of these spring

Figure 3.32

scales lined up alongside one another allows us to compare the weights of more than two items, which cannot be easily done with a set of balance scales.

To prepare for a broadening into number, we can introduce a non-standard, fairly uniform unit, such as conkers, pebbles, glass beads or cubes, which can be used to measure the weight of the items, in a similar way as described in the measuring tapes example, above. For example, we can explore *how many conkers* they think will 'stretch the same distance as' a potato compared to how many it takes to equal the weight of a

Figure 3.33

glue stick. Alternatively, by keeping the item to be compared constant (the potato) and providing a range of units, we begin to focus on the impact the size of a unit has on the count.

If we accept that the principally number-based mathematics within school often contrasts markedly to mathematics learning outside school, we might focus more on the rich measures experiences we can provide that feed into children's developing number understanding.

Figure 3.34   Homemade spring balance.

PREPARING FOR WORK ON NUMBER PLAY

Playful number learning involves providing authentic problems to solve and leaving space for children to work out their own solutions whilst building in time to discuss these afterwards. It involves being alert to the opportunities that children's free play presents and

Figure 3.35

being flexible in how to take these forward, by becoming knowledgeable of the possible trajectories of learning and development in number. The work of Clements and Sarama on mathematical learning trajectories (2021) is indispensable for this. Their website, which requires registration but is free, outlines developmental trajectories for all areas of mathematics: **www.learningtrajectories.org**. Trajectories of learning are discussed further in Chapter 4 ('Adults – What do we do?'). Here I describe three examples of playful number activity.

EXAMPLE: PIRATES' GOLD

Figure 3.36

On a tablecloth on the ground I place a collection of gold coins. There are two toys sitting on the cloth wearing pirate hats. Each pirate wants their fair share of the gold. The problem posed is, can they each have a fair share?

How have I prepared this? I consider the number of coins to provide. It has to be on the edge of these children's counting range (maybe 12 is a good choice). Do I want an amount that shares equally (probably to begin with) and/or one (later) that does not? I provide clip boards and pens as I want every child to show a solution on paper for us to compare and discuss as a group. I am ready to introduce a third pirate at an appropriate moment. I ponder how my children might approach solving this problem and what I might look out for. Maybe they might 'deal out' gold coins one at a time? Maybe they might give each pirate toy a handful. Do children check that each toy has a fair share? Do they do this by counting?

This problem comes from the research of Davis and Pepper (1992). Sharing is a problem-solving context that makes complete sense to young children. We can build it around a story, we can change the amount of gold, or change the number of pirates and explore (play with) what happens. Sharing 12 items between two toys and introducing a third and then fourth toy is a mathematically rich problem as the amount 'works' well (12 has 2, 3, 4 and 6 as its factors), but maybe on another occasion I want them to find out what happens if we attempt to share a prime number, such as 13. Davis and Pepper (1992) found this problem led to some high-quality reasoning amongst pre-school children and that even young children can invent creative ways of solving the problem where an amount of biscuits did not share fairly between dolls: *Give him an extra one 'cos he's hungry, Break that one into crumbs.* By encouraging children to solve the problem on paper, we add an extra layer of thinking which provides us with a range of solutions for them to explain to us after the event. As I said in Chapter 2 ('Fostering mathematical reasoning'), real life moves too fast for young children to reflect on the mathematics they have been doing at the time.

EXAMPLE: GEMMA'S SHAPES

In this example, a child has chosen to sit at a table 'laid' with a variety of regular and irregular flat shapes in a range of sizes and counters, paper and pens. I watched her carefully selecting a shape and drawing around it. She filled her outline with teddy-bear counters. She repeated this with different shapes. After a while, I might ask: *What have you noticed?* or even, *Which shape*

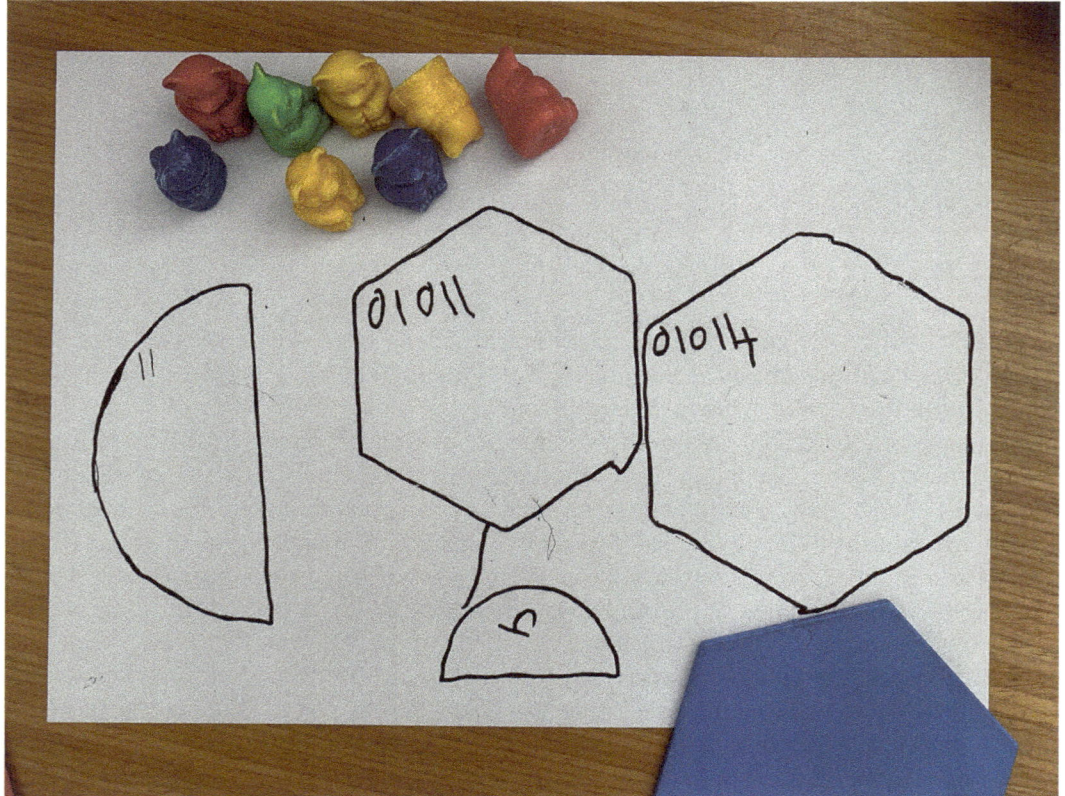

Figure 3.37 Gemma's recording.

fits the most teddies inside? She chose a large hexagon from the tray, drew around it, filled it with bears, counted the bears by moving them one by one outside the outline and said *ten, ten, one.* I said nothing and handed her a pencil. She wrote: '0 1 0 1 1', writing each ten from right to left. She then continued to draw around different shapes, fill, count and record in a similar way.

How do I interact on this occasion? It would be very easy for me to instantly correct Gemma's recording and show her how to write 21 or 24 any of the numbers from 20 onwards. Indeed, this will be a lovely future teaching session with a group looking at a 100-square, counting aloud, writing numbers and discussing the patterns. But I do not have to do this now; instead, I enjoy her playing with how larger numbers are made up; two tens and some ones. I probe a little into what she is seeing and doing. It is a remarkable insight into her understanding of number. Being playful means that rather than rushing to correct, we register what we notice for later, when we can prepare something to explore the issue in some depth, and we can continue to share the child's current mathematical engagement and interest. This is being prepared for learning to take place over, rather than within, time.

EXAMPLE: PLAY WITH CALCULATORS

Calculators have had a rocky ride in schools in England and yet they can bring a lot of numerical joy and wonder to children even as young as 3 years of age. For them a calculator is a toy, but one that embodies our number system. Leaving some calculators out with paper and pencils, encouraging children to record the numbers they find, then intermittently observing what plays out is very informative. In this section are some examples from children of 4, 5 and 6 years of age.

Children often begin by making numbers familiar to them appear in the display. You can see examples here of meaningful numbers as *labels* rather than quantities, which is the most common way young children encounter numerals in the world outside schools and settings; here we have door numbers, telephone numbers, bus numbers and the like. These are important to them and I can encourage them to think about which key to tap first, next, etc. in order that the number appears correctly in the display. We can discuss how, for instance, a '4' appears in different formats. Introducing 'calculator booklets', or a floor book, where children freely write and draw as they play, is a great addition as children's explorations develop and chart a child's interests and their growth of knowledge over time. Moreover, inviting children to write 'their' numbers, such as telephone numbers, requires them to read and record the numerals correctly from left to right. Calculators can inspire an interest in longer – and larger – numbers, as children ask for a full display of numerals to be read aloud (you become very adept at this). The wonder they show as we read 317853021 as: *three hundred and seventeen million, eight hundred and fifty-three thousand and twenty-one* is a joy to behold. As they play around with this idea, we can observe how a child gradually moves towards a realisation that 999,999,999 is the largest number we can display on a simple calculator, or that something followed by six zeros indicates 'something million'. We can introduce the 'constant function' in order to generate the counting sequence. By modelling pressing: '+, 1, =, =, =, ...' repeatedly, the calculator 'counts' up from 1 in the display. If we start with a larger number and press '−, 1, =, =,' repeatedly, it counts down. I have heard children refer to them as 'countulators' when exploring this function. In Figures 3.41 and 3.42, some 5- and 6-year-old children are playing with counting sequences going up in steps other than one, by pressing '+, 2, =, =, =,' and '+. 100, =, =, ='.

(Note: on some calculators we have to press the function key + or – twice to activate the constant function.)

In preparing for this play I have made myself aware of how play with a calculator might link to my children's wider developing number understanding. This is play that supports children's understanding of the pattern of the number sequence, ordinality (the attribute of number indicating position and order), facility with counting from and to different numbers and in different steps, and familiarity with how numbers are written and read. All of this is valuable number knowledge for children to develop, as understanding ordinality has an impact on later mathematical development (Haylock and Cockburn 2017).

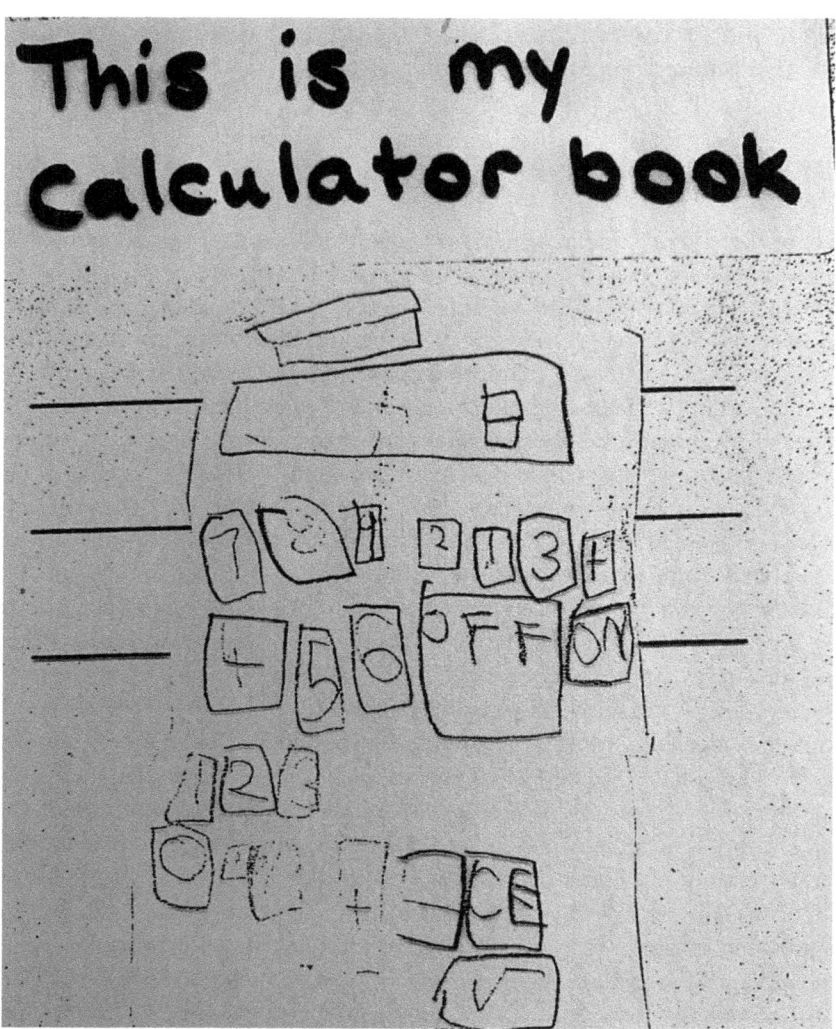

Figure 3.38

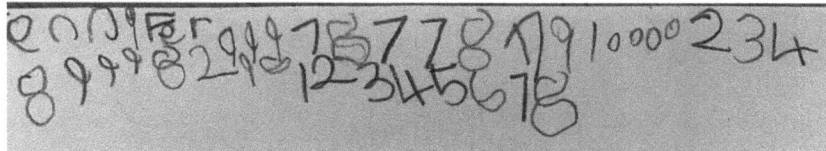

Figure 3.39

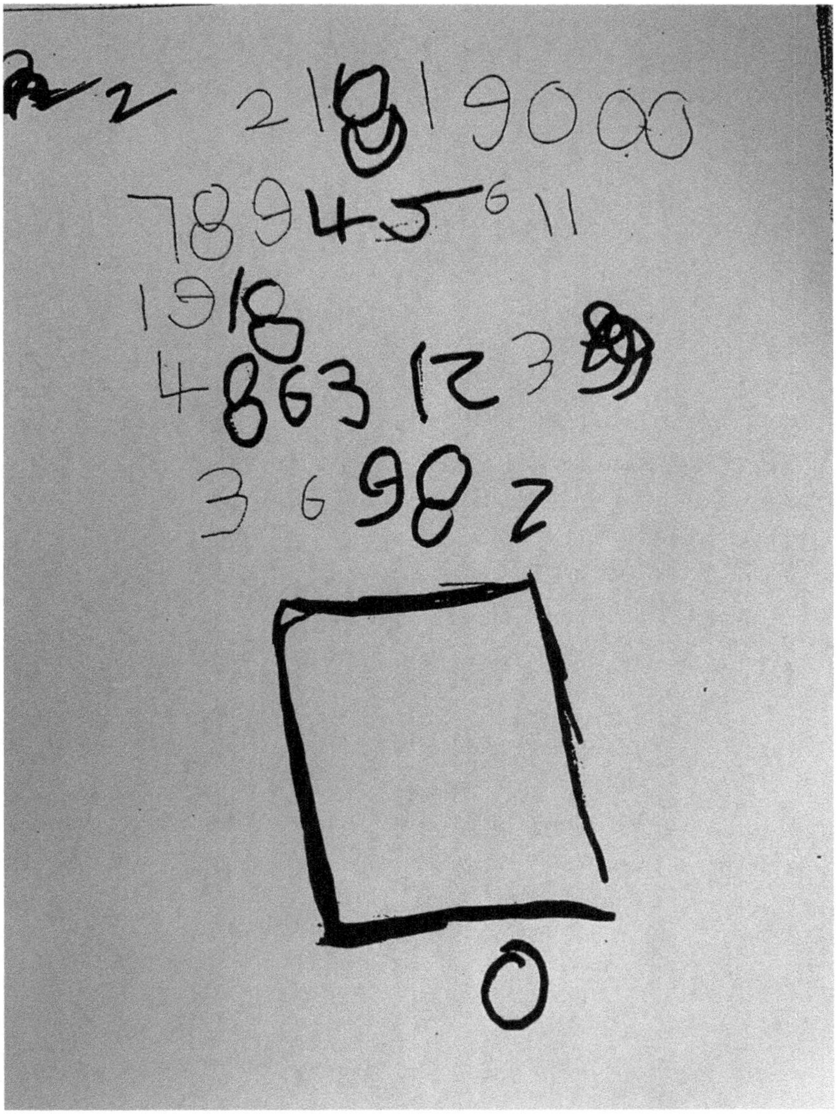

Figure 3.40

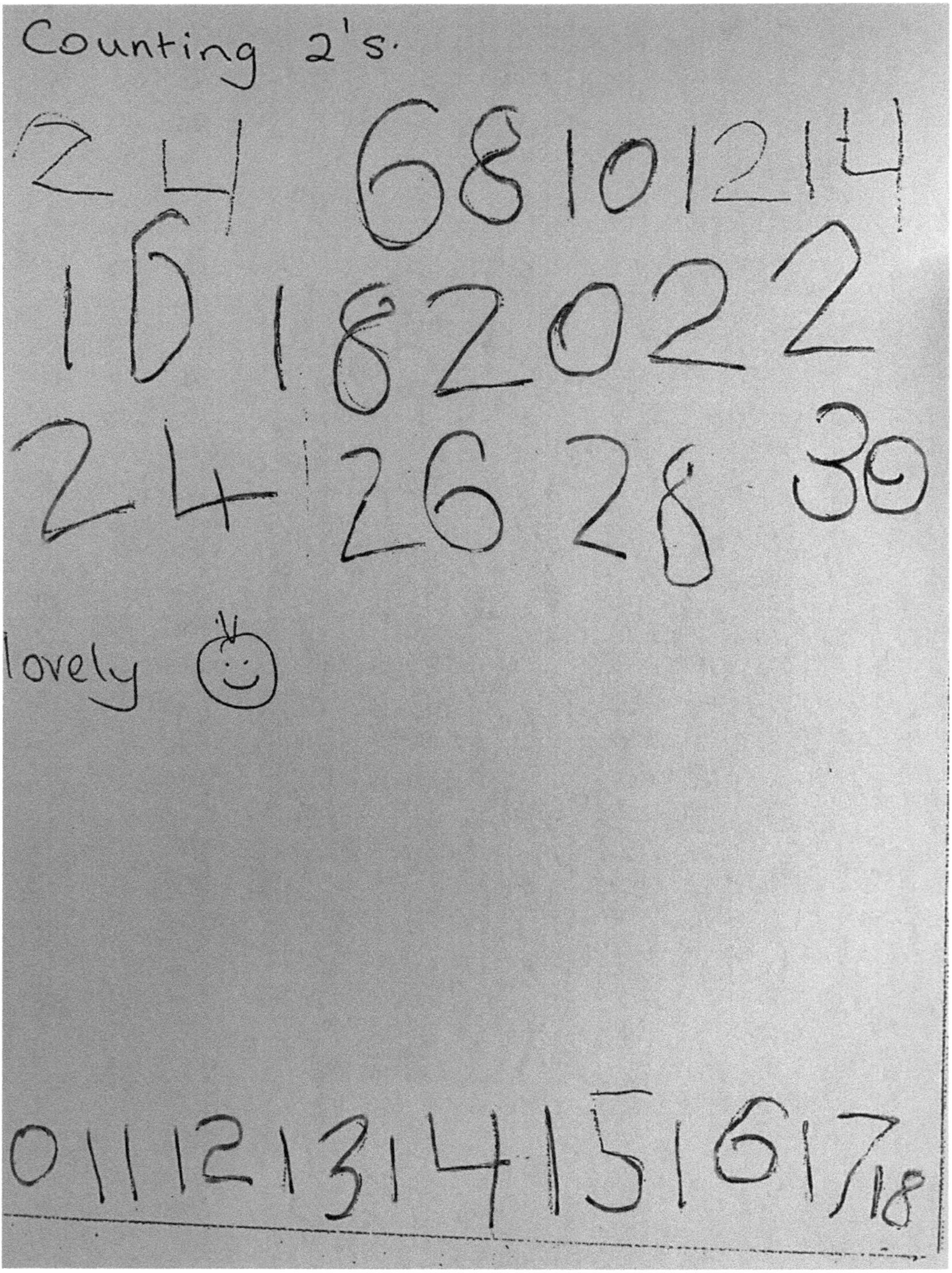

Counting 2's.

2 4 6 8 10 12 14
16 18 20 22
24 26 28 30

lovely 😊

0 11 12 13 14 15 16 17 18

Figure 3.41

Figure 3.42

PREPARING WITH GAMES IN MIND

I honestly believe that most of the Early Years mathematics curriculum could be approached through games. Simple games with few rules and lots of opportunity to vary these are the most powerful. By sharing a game with families, they can also provide a positive, unthreatening link between home and school. When a game becomes familiar, children playing as partners against another pair can be very powerful, as they are encouraged to discuss and agree what to do each turn. Games provide an important opportunity to practise skills that require a lot of repetition to achieve fluency. At this stage counting accurately is a key teaching focus, and many counting tasks can be transformed into a game by the simple addition of dice. In the following example this is exactly what happens and Alice and Sasha (both aged just 4 and in a Nursery class) invent a collecting game whilst playing with a pile of pennies, some small pots with lids and dotty dice.

VIGNETTE: THE COIN GAME

Figure 3.43

Figure 3.44

Franchesca: I've got much more than you.

Alice:	No, that's not how we play. You have to put them all there and you have to shake the dice and see which number it is and you take that many coins, how many you get.
Sasha:	So, we have to go, one, one, one! You've just gone all wrong!
Alice:	I'll start first. Which dice do you want, Franchesca?
Sasha:	Mine's going to be ... I've got ... how many?
Alice:	You've got one, put one coin.
Sasha:	One in there.
Alice:	Franchesca hasn't had a go. That's three. One, two, three. You've got to shake the dice.
Sasha:	What's that number, Alice? What's that number?
Alice:	(*Looking hard at her own dice*) That's that number. I've got to pick it right up to the end of 30! Franchesca, it's your turn, 'cos I've just won a lot actually!

Figure 3.45

Figure 3.46

Games are cultural and children have to be introduced to this culture before the mathematics becomes either obvious or useful. It is worth remembering that most babies are familiar with the rules of variations of hiding games like 'peek-a-boo' and hiding and revealing is a useful starting point for a mathematical game: *What do you notice? What is the same? What has changed?* If younger and less-experienced children are encouraged to play games in their own ways, they will engage. Gifford (2006) has written about the value and practicalities of games: in short, they offer opportunities for peer learning, problem solving, reasoning and discussion, and importantly *protect children's self-esteem, as the element of chance shifts the balance of power away from the adult* (Gifford 2006: 50).

How have I *prepared* for the coin game to take place? To begin with, or with younger children, games without rules work well, such as the example above. I have thought carefully about the resources to select, paying attention to the size of the containers I provide (large enough to hold enough coins to be interesting but not too large to be outside the children's counting experiences). I used a large amount of 1p coins to attract interest and because they are all similar to avoid distraction of colour or type, and added a collection of dice to imply a game. On another occasion, I might extend this game and use different items

Figure 3.47

from a familiar treasure or loose-parts collection, to look at similar quantities of differently sized objects, or add some chocolate-tray inserts or ice-cube trays to separate the counting items and look at *how many more to make* ...

To invite children to invent their own games, add large dice to outdoor grids, tracks or stepping stones and build in opportunities to discuss with the children the games that emerge, creating rules collaboratively. Discuss with the children the games they already play and are familiar with. Mix and match the dice used; numeral dice, dice that show '0', and as well as dotty-6 dice, also use dice with fewer spots – up to four to encourage subitising (the instant recognition of small amounts) and model how to roll the dice from a cup, looking at the number rolled and covering the dice before saying the amount out loud.

FOUR SIMPLE GAMES

The following four games work well, with adjustments for age and experience, with children between the ages of 3 and 6:

- 'Stick race' is a measurement game.
- 'One or two' and 'Pass it on' are both number games.
- 'Make it the same' is a spatial game.

All involve strategy and all can be easily adapted to suit more experienced or older play-ers. It is more difficult to ensure games are 'low-threshold' enough to be accessible for *every* child, and that children take control of the game rather than you. The trick is to find one game you all enjoy and to repeat it over and over. Children have favourite stories and books they like to hear repeatedly; we would also like them to have favourite games they want to repeat. When the mechanics of a game are familiar, we can begin to discuss the mathematics, what happened and why, and what is likely to happen; we open up opportunities to work on strategy, to predict and reason by asking: *What might happen if we change the dice/the containers? I wonder why you think that? What might you do differently next time you play?*

All these games are for two or more players and can be taught with an adult as one player and a child, or group of children, as the second player. Later, as a game becomes popular and familiar, children can play with each other without an adult.

Stick race: a length game

Figure 3.48

Figure 3.49

You need:

- to collect a pile of all sorts of length of stick;
- chalk or similar to make a mark on the ground.

TO PLAY

- Mark a 'Start' line on the ground. Now mark a 'Finish' line. Collaboratively, you are going to make a line of sticks long enough to reach the Finish line.
- Decide who starts.
- The first person lays a stick with one end on the Start line.
- Take it in turns to choose a stick and lay it down so one end touches the end of the last stick laid. Keep taking it in turns until you reach the Finish. The object is to choose your sticks carefully to be the person to lay the stick that is as close to the Finish line as possible, maybe even touching it.

MATHS DISCUSSION

- How close did you get to the finish? Are all your sticks touching end to end? Why is that important? How are you choosing your sticks each time?
- How many sticks did it take this time, to reach from Start to Finish?
- Play again.

WHAT HAPPENS IF?

- You change the Start and Finish lines?
- You each lay your own line of sticks?
- You don't have to lay a *straight* line of sticks?
- What about playing with Duplo bricks?
- What if more people play? What will you try?

OLDER PLAYERS

Principally we will be exploring how the size of the unit affects the amount we require to fill a space (see page 80 on measures). To develop this idea, we need to play with non-standard units that are all equal or uniform (rather than selecting sticks of different lengths), such as similar bricks or loose parts of similar size.

Figure 3.50

You can choose to either:

- fix the distance: each player chooses a different unit to make their own line:
  - *You are racing with bottle tops, and she is using bricks, I wonder who will touch the finish exactly?*
  - *How many of these bottle tops do you think will reach the finish if 20 bricks do?*
- change the distance and keep the unit constant:
  - *If it takes 20 tops to reach from here to here, how many do you think will it take as far as over there?*

THE MATHEMATICS IN THIS GAME MAY INCLUDE:

Recognising and comparing different lengths, matching and estimating length/distance, aligning objects with precision, counting accurately, selecting lengths to match a visual distance, making decisions, predicting, visualisation, enjoyment.

One or two: a number game

You need:

- a dish of some sort (or draw a 'dish' on paper);
- 12 items, such as buttons to place in it.

Count the items into the dish together to establish there are 12.

TO PLAY

Decide who starts. When it is your turn, you can choose to remove one or two items from the dish.

The person who removes the last item loses.

MATHS DISCUSSION

Play again (and again, and again). Does it matter who starts?

Can you make your opponent lose? How?

WHAT HAPPENS IF?

- The person removing the last item is the winner?
- You start with fewer items?
- More people play?
- If you choose to remove one, two or *three* items each time?

Figure 3.51

OLDER PLAYERS

This game does not require much alteration; extending it lies in developing a strategy to *always* win, with whatever starting number and with any rules.

'Nim' could be a development, a game said to originate from China. You begin with lines of tally marks or sticks, such as shown in Figure 3.52.

Players take it in turns to cross out (or pick up) any number of tallies (sticks) provided they all come from the same line. Depending on the version being played, the goal of the game is either to avoid crossing out the last tally, or to cross out the last tally.

Figure 3.52

THE MATHEMATICS IN THIS GAME MAY INCLUDE:

Counting accurately, checking and comparing quantities, algebraic thinking such as recognising and analysing patterns, understanding why something happened, and anticipating and predicting what will happen (several goes in advance), making decisions, making generalisations, analysing how things change, explaining a winning strategy, enjoyment, resilience, perseverance.

Pass it on: a number game without a winner

You need:

- a collection of 'treasure' or loose parts of any size;
- and for each player: a dish or suitable container, a dice and cup to shake this from.

TO PLAY

Each player selects some treasure for their dish. Decide who starts. When it is your turn, roll your dice and pass on this amount of treasure items from your dish to another player.

MATHS DISCUSSION

What happened? What will happen now if you roll a 3? What do you *wish* to roll next? Why? What if … you haven't enough treasure to pass on? Did anyone win? How did you decide who was the winner?

WHAT HAPPENS IF?

- The person who ends up with an empty plate is the winner?
- You each begin with exactly five items?
- You have to pass it on to your left?
- You play with different 'treasure', e.g. sand?

Figure 3.53

OLDER PLAYERS

This game has such a low threshold, once played and understood, it can be developed in many directions, according to your mathematical focus. Using egg boxes, five- or ten-frames and items that fit into these, and watching them fill and empty as the dice are rolled is an ideal opportunity to look at *how many more to make …*

We could introduce a start or end goal, where we all start with ten things, or we end the game when someone has filled two ten-frames.

With older children perhaps we could pass on 2p and 5p coins and look at amounts collected, or measure out water and see who can reach exactly 500ml. The key in being playful is that you and the children, being familiar with the structure of the game, come up with adaptions to try collaboratively.

THE MATHEMATICS IN THIS GAME MAY INCLUDE:

making decisions about how to play effectively, subitising (using a dot die), recognising numerals and relating to quantities (using a numeral die), knowing number words in order, counting small amounts, accurately, tagging each (dissimilar) item with one spoken numeral, accurately counting a small amount from a larger amount, understanding the cardinal value of small amounts, checking and comparing small quantities, algebraic thinking such as understanding why and how a quantity has changed and anticipating and predicting what might happen if (I roll a zero, for example), enjoyment and control.

Make it the same: a spatial game

You need:

- two identical sets of items, one for each player, for example;
- about eight to ten blocks or bricks of different shapes;
- a couple of play-people, a vehicle and a tray or sheet of card each.

TO PLAY

Both players make sure they each have exactly the same toys and bricks. Decide who starts. This player (the builder) builds a small-world arrangement using all the toys on their tray. When they are satisfied this is complete, they call the other player (the copier) who makes an identical arrangement, with player 1 helping by giving instructions.

When they are satisfied the two arrangements are the same, they call out, *The same!*

Swap roles and play again.

MATHS DISCUSSION

What does *the same* mean? What do we mean by *exactly the same? nearly the same?*

What did you hear [Tom] say that helped? What didn't help? What would have helped?

What happens if?

- You change the items?
- You play with some Lego?
- More people play?
- You stand up a barrier (a hard-back book)

Figure 3.54

for the builder to sit behind with their collection, making their arrangement secretly. The builder gives directions to the copier but only the builder can peep over the barrier to check how the copier is getting along, trying not to touch or point. When the builder is satisfied the two arrangements are the same, the barrier is removed for comparison.

OLDER PLAYERS

Again, this is a low-threshold game that can become challenging quite quickly. You might choose to begin with a builder and copier laying brick by brick. The object is for *all* children to enjoy exploring the relationships between shapes, rotations and translations, and visualising and describing different arrangements. You could try giving a few quick glimpses of a completed small model or arrangement built that is then covered, with the children making a replica from memory and maybe more glimpses.

Playing with bricks that are all the same colour makes this more demanding, or with items all the same shape such as interlocking cubes, as does using materials that make more abstract patterns, such as pattern blocks.

THE MATHEMATICS IN THIS GAME MAY INCLUDE:

Decision making, visualising spatially what something might look like (if rotated, flipped, transformed, etc.), understanding relationships between shapes, recognising (large and small) similarities and differences in spatial arrangements, using accurate descriptive language, anticipating what a partner might do, enjoyment, control.

TO SUMMARISE

In order to support all children's mathematical learning, we need to provide both time and space for immersion in play and time to discuss and review that play. We need to plan time for both open-ended play and goal-oriented tasks, as each benefits the other. Careful selection of games that all children can access can help us focus on our children's mathematical well-being, that is, their confidence and enjoyment in mathematics, as well as allowing us to observe what they know and can apply in another context.

Preparation has three main principles:

- leaving plenty of space in our planning; it really is not possible to plan beyond Wednesday in any one week if I aim to be responsive to what the children are learning;
- being flexible and expecting surprises, enjoying the threads of investigation the children begin;
- researching the possible developmental trajectories in the areas of mathematics I am focusing on in order that I am clear how I can best develop the mathematics that plays out;
- planning time with the children as a group, to review and discuss what I have observed and what they have been engaged in. These reviews can be as informal as suits your style and situation, but frequent. They are opportunities for adults to stimulate discussion about, and reflection on, the mathematics, and to both assess and move children's thinking on.

In the following chapter, I explore in more depth the role of the adult working mathematically with young children and take a look at developmental trajectories in mathematics.

4

ADULTS – WHAT DO WE DO?

Carer to child, after her first day in school:

Did you enjoy it? What did you do today?

Child: Well, I spent a long time sorting out Mrs B's animals, but when I finished, she tipped them all back in the box.

SOME BACKGROUND

This child (the exchange did take place) was bemused that she had been asked to complete a task that had so little value. Where is the human sense in sorting something out and then tipping it all back again? What impression of school and of mathematics and its relationship with the outside world did this leave on the child? This chapter will explore ways of working mathematically in partnership with children. It analyses the role of the adult in young children's mathematics learning: what we teach and how we teach effectively.

Many adults do not feel as confident mathematically as they do in other areas of learning. For many adults, mathematics has been a largely negative experience:

It is well known that there is wide variation in attitudes toward mathematics, and that many people have extremely negative attitudes toward mathematics, often involving strong emotions of fear and anxiety.

(Dowker 2021: 4)

This is compounded in the UK by it being culturally acceptable to express disinclination and incapacity in terms of mathematics (National Numeracy 2019). It is particularly unhelpful for families to say to children: *Don't worry, I was never any good at maths*. It is quite staggering to me that evidence from 20 years plus illustrates we have not managed to change much in relation to this. Evidence from both the 1999 and 2019 Trends in International Mathematics and Science Studies (TIMSS) comparing pupils' mathematical competence at 10 and 14 years of age over 40 countries showed students' attitudes to mathematics as continuing to decline over their school years (TIMSS 1999; Walls 2009), with attitudinal questionnaires to pupils and teachers showing a wide gap in achievement between those having confidence in their ability and those with much less confidence (TIMMS 2019), so we know attitude and confidence to be crucial and yet do not appear to be doing much to alter this. Interestingly, Askew et al. (2010) found that so-called 'high-performing' countries in terms of international comparisons were as concerned about students' negative attitudes to the subject as those countries that were seen as lower-performing. It seems our attitude to mathematics is not solely dependent upon being 'good' at it, but a complex combination of emotion, belief and social context.

How we work with children, how we respond to them, as well as what we teach, is critical in affecting children's attitudes. How we as adults working with young learners help our children develop into confident as well as competent mathematicians is the focus of this chapter. Moreover, it is as important that the mathematics we do with our children is engaging for the adult as it is for the child, so we will look at that as well. Much of what is said in this chapter applies to learners of all ages and at any stage of their mathematics education. Mike Ollerton is an experienced secondary mathematics teacher:

> In general, deepening learning is about developing a questioning-by-learners culture where learners feel safe, where they are strongly encouraged to ask questions and offer conjectures.

> (Ollerton, in conversation, 2020)

Here, Ollerton makes links between the nature of mathematics (as solving problems), how we learn mathematics (by exploring and questioning) and the kind of environment needed (safe enough to explore and question) as well as how we produce this (by openly valuing and supporting such explorations). This book is clear that the role of all adults working with children contributes fundamentally to the environment we establish for mathematics learning, and is as important when interacting in everyday experiences, in continuous provision and child-led activity, as in moments of direct teaching. We:

- make ourselves aware of the mathematical potential around us and in what we offer. Robertson (2017) refers to this as asking ourselves: *Where's the maths in this?*
- provide all kinds of interesting resources, tasks and tools, indoors and out;
- listen and observe with care, assessing what children know and what they bring to the tasks;

- encourage children's curiosity by showing real interest in their questions and investigations that result;
- educate ourselves in the developmental trajectories of various areas of mathematics.

These elements are covered under the following four sub-headings:

- **Tuning in:** exploring the environment and culture surrounding the teaching of mathematics.
- **Balancing teaching and play:** examining how explicit teaching episodes interrelate with freely chosen play activities.
- **Big Ideas:** reflecting on the Early Years curriculum; the 'What', knowledge of learning trajectories and the 'How', the Characteristics of Effective Teaching and Learning (Department for Education 2021).
- **Conversations:** considering our interactions with children and the role of 'sustained, shared thinking' (Siraj-Blatchford et al. 2002).

Figure 4.1

TUNING IN

Children are naturally curious, exploratory and imaginative, so how might we foster this and make use of it in our mathematics teaching? The learning environment consists of the *experiences* and *spaces* both outside and in, as well as our *interactions*, and it is helpful to view the environment through the eyes of the children we are working with. Consider the following four questions:

ARE THERE A RANGE OF MATERIALS TO EXAMINE AND EXPLORE OVER TIME THAT ARE MATHEMATICALLY INTERESTING?

These should not be confined to a 'maths area' and need to include natural and mathematically structured items, including piles of things to feel and count (loose parts of all sizes), containers of different sizes to fill and empty, blocks of all shapes and sizes, jigsaw puzzles and mosaic-type tiles and pattern blocks for pattern-making. Materials children can see as well as handle, imagined reality, games, puzzles, 100-squares, number tracks, games and paradoxes are fascinating to children. Something that intrigues might be mathematically structured, such as a tub of pattern blocks or a calculator, or something natural like a sunflower head full of seeds. Questions that I can observe arising are:

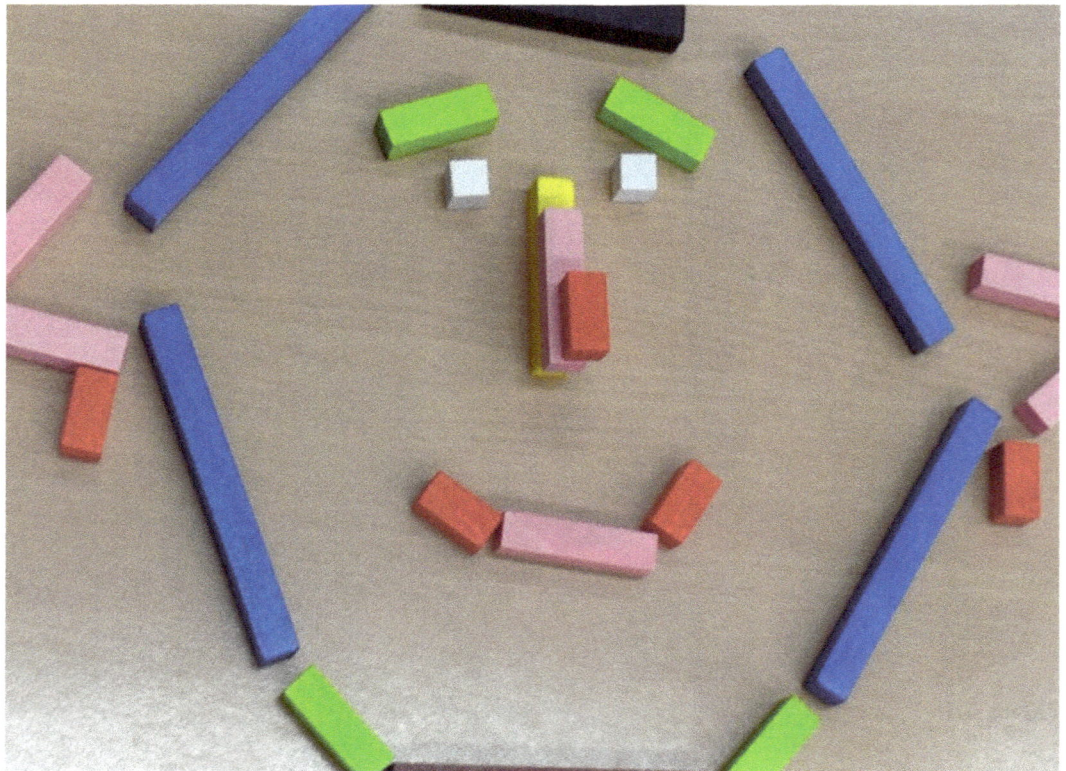

Figure 4.2

Figure 4.3

What is that long number shown on the calculator?

Why can't I roll a six every time?

What hole does this shape fit into?

Is that leaf the biggest one I can find?

Under which cup is the coin hiding?

How can 50 things fit into a tiny box?

Why are my feet size 11 and yours are 5, and yet mine are smaller?

Figure 4.4

What we notice children doing, as well as longer observations of how children interact, will help us shape appropriate mathematical tasks to try.

Figure 4.5

CAN CHILDREN ACCESS RESOURCES INDEPENDENTLY?

Tidying away or 'stocktaking' can present rich and repeated opportunities for counting and matching to check and correct.

(Davenall and Gifford 2015)

Figure 4.6

Regular opportunities to organise shapes to fit, deciding what goes where, and checking amounts is time well-spent, offering regular openings for mathematical reasoning.

ARE THERE OPPORTUNITIES FOR CHILDREN TO DEVELOP THEIR MATHEMATICS PLAY OUTDOORS AS WELL AS INDOORS?

Being outdoors plays a unique role in children's development:

> *The natural and built worlds provide dynamic and constantly changing environments, offering an endless supply of patterns, textures, colours, quantities and other attributes that underpin much of the necessary early maths experiences.*

> (Robertson 2017: 1)

Figure 4.7

Robertson gives many examples of how to capitalise on the outdoors mathematically. For example, in mud kitchens we can mix potions of different strengths, spaces can be built and explored, many table-top track and collecting games can be adjusted and played 'huge and outdoors' (to paraphrase Gifford 2004).

ARE WE ABLE TO SPEND TIME TUNING-IN TO OUR CHILDREN?

How we interact with young children is a function of our wider relationships. Tuning in involves observing what children are doing over time and listening carefully to what the child is saying, with their body as well as their voice. *What do you see?* and later,

Figure 4.8

What do you notice? are useful opening questions when interacting mathematically with children; it gives us time to tune in. Our body language and tone give important messages. Research into effective Early Years mathematics teaching is extremely clear that there needs to be a blend of adult instruction and play-based, child-led activity, to effectively support mathematical, as well as communication and language, development (Anthony and Walshaw 2007; Clements and Sarama 2021; Downton et al. 2020; Pascal et al. 2019; Siraj-Blatchford et al. 2002). Adult-initiated sessions are invitations for children to enter the world of the adult and child-directed mathematics is a necessary complement to this, where we enter the world of the child. Teaching needs to be both *responsive* and *intentional*. It is difficult to see how this can be achieved effectively without adequate time for observation.

In the vignette 'Counting bumps', a young child makes a conjecture: *Can you have a triangle number?* and noticing this comment presented the adult with the opportunity to intentionally work on noticing, describing and predicting patterns. I could stimulate this by re-capping what they say and waiting for their response, for example: *So you think that we can make some triangle patterns?* By tuning in to children I am suggesting we make time to notice children's interests that could lead to further mathematical experiences. This is being *responsive* to what we observe. Through this, some *intentional* teaching can take

VIGNETTE: COUNTING BUMPS

Figure 4.9

We were counting the bumps on the Duplo bricks in a feely bag and one of the children (4 years of age) said: You know the squares, because it's the same dots along and the same up. And I thought: That will help your understanding of square numbers later on. I mentioned square numbers and they wanted to explore what other shapes they could make with dots, asking: Can you have a triangle number? We used corks to print, paint and count triangle numbers.

(Stacey Anderson, Reception teacher
@Theplayacademy, in conversation, 2020)

place, through which we can inspire more than that one child to investigate further. This need not happen in the moment we notice something. I might be surprised when children take an activity in an unexpected direction – making huge pieces of fruit with the interlocking cubes ('pears') instead of my intended 'pairs', for example – but maybe I give myself time to devise a task that taps into this interest – comparing the weights of the 'fruit' constructions, perhaps. We should not underestimate children's capabilities, nor be afraid of challenging tasks. By 'challenging tasks' I mean tasks that challenge children to think, rather than increasing 'difficulty'. When 'tuned in', we can enjoy the directions that children take us and become engaged ourselves, as adults. When something comes up that I feel ill prepared for I can 'park' it, by saying something like, *Wow, that's interesting. I hadn't thought of that. I am going to jot that down and have a think about that. Let's chat about it again tomorrow/later.*

Figure 4.10

Whitebread's research has consistently demonstrated superior learning and motivation arising from playful approaches to learning in young children (Whitebread et al. 2012). Johnson and colleagues (2019) make the point that many children in their study demonstrated understandings of counting principles that were not captured by simpler tasks, when they were asked to complete more complex tasks. If the task has a 'low-threshold' entry point, makes sense and is intriguing, young children will become involved and may well operate at a higher level of thinking and we can help them develop their thinking further. Two things interrelate and are important here:

- firstly, making ourselves aware of the mathematical potential, the trajectory of the mathematics (see later, under 'The Big Ideas');
- and secondly, the *culture of learning* we establish within our setting for mathematics.

If the culture of the setting is one where learners feel safe, where they are strongly encouraged to ask their questions and offer their conjectures, in mathematics as in any area of learning, they often reveal understandings beyond those we might expect. If I routinely ask: *What do you notice?* I will receive a range of responses which we can talk about and children will learn that all contributions are equally valued and all add towards our mathematical

Figure 4.11

understandings. It positions children as authoritative and capable, respecting their interests and them as emerging mathematicians. Klein (2007) has argued that often classroom practices, in mathematics in particular, counteract any perception of learners as authoritative and capable, working against the production of confident and proficient young mathematicians and consequently turning learners off mathematics. This is crucial to consider if we wish to change the dominant culture surrounding mathematics as difficult and only accessible to some. How we answer children's ideas and responses defines how they feel about themselves in relation to mathematics – in short, the mathematical identities they develop (Askew 2008).

The sharing of authority with young children over how the mathematics plays out is tied up with what I have referred to as 'preparation rather than planning' (see Chapter 3, 'Preparation rather than planning'). Here are two ways of tuning in mathematically:

- We notice something that the children say or do that can lead to some mathematics (as in the 'Counting bumps' vignette).
- We find something we think will interest our children that has mathematical potential, maybe a short video, image or resource, and use this to set the scene for and hook them into some mathematics, that could develop in several directions (see an example in the following vignette, 'Dotty cards').

VIGNETTE: DOTTY CARDS

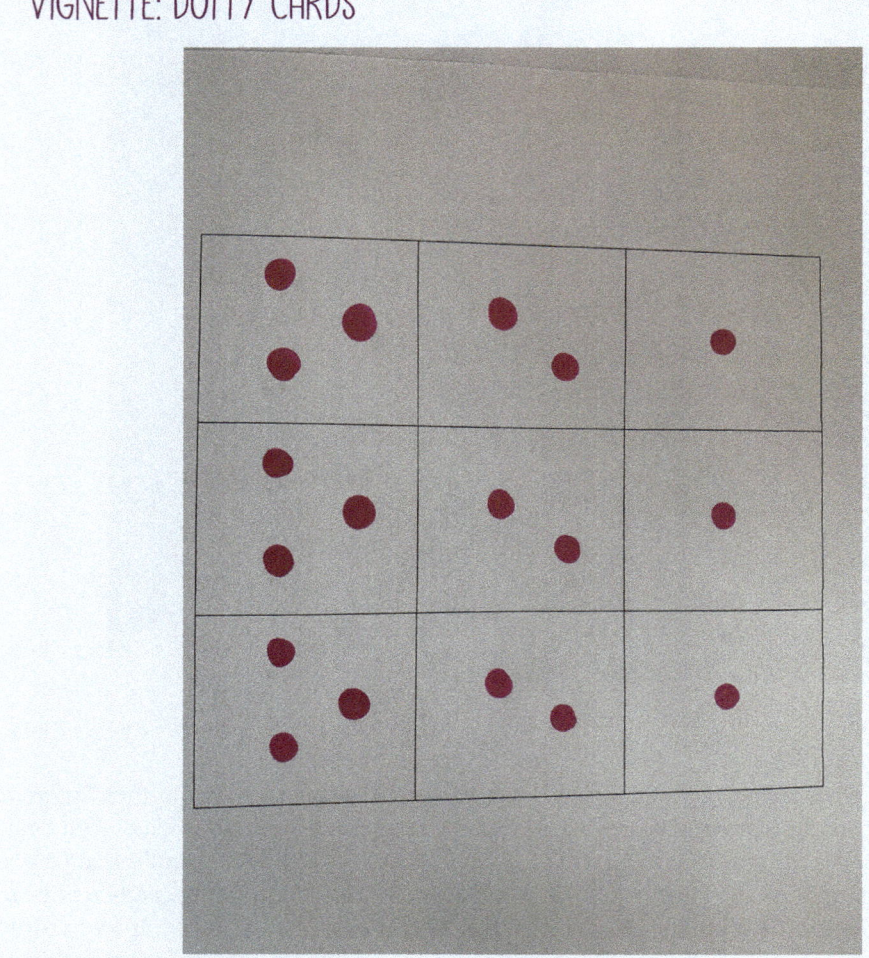

Figure 4.12

Figure 4.13

Bird (1991) describes a task where she gives each child a set of nine small square cards with one to three dots on them, as shown. After reading about this in Marion Bird's book I have enjoyed this task numerous times with children between the ages of 3 and 6 years. As they examine their cards, I ask: *What do you notice?* and listen to the various responses: *dots, squares, ones and twos, all threes, the same, three, three, three again!*, and so on. I observe children beginning to sort them into cards with the same number of dots. Whilst playing with the cards, they are getting to grips with the structure of the set. Can they spot when one of the set of cards is removed? I can invite them to *elaborate* on what they are seeing, e.g. *I really want to know more about this pattern …*

(Continued)

Next, more likely on a later occasion, providing 3×3 grids on to which the cards fit further structures what they see. I observe what each child does. Some lay the one-dot cards horizontally on to the grid; some work vertically, starting with the three-dot cards and 'counting down' to one. Some copy the dots on to the grid. They are always eager to describe what they have done.

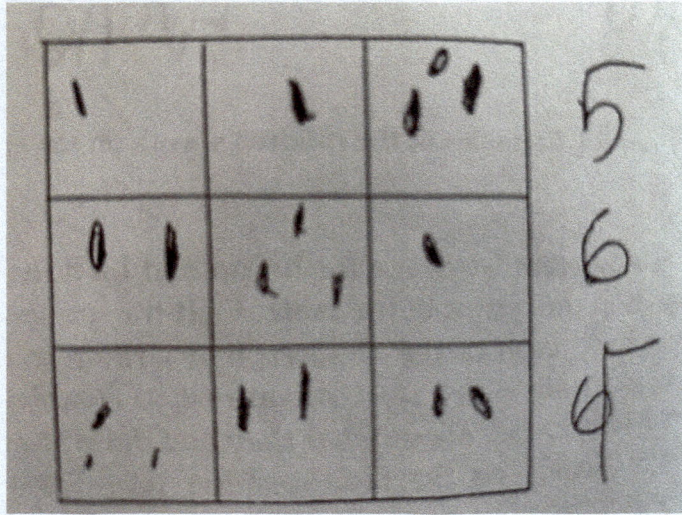

Figure 4.14

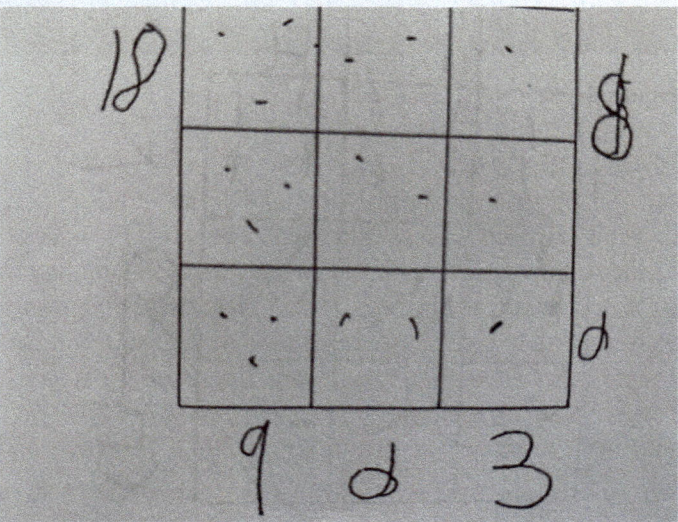

Figure 4.15

I can choose when, and if, to ask the next question I have up my sleeve: *How many dots can you see altogether in this row?* We continue to look at one grid and the number of dots in each row, writing the totals, and then I can suggest they might count the dots in the rows in their own grids. This is where I choose to leave them, with plenty of grids to play with. The children explore making different row and column totals, recording if and how they wish. Bird describes one child becoming excited at making six in every row and I have seen many children spotting this pattern, as well as other patterns, in their own grids.

Both noticing and seeking out something depend upon planning for *foreseen possibilities* (Bird 1991: 133). Here we consider what *might* happen when planning an activity, or suite of activities, but do not let that restrict what happens in practice (see Chapter 3, 'Preparation rather than planning').

The small totals attained by adding the rows and columns in the task outlined in the above vignette are perfect for this age range, with the small amounts on the cards encouraging perceptual subitising. This task is mathematically well pitched for this age of child, with plenty of mathematical possibilities to explore together.

Later, providing blank cards for children to make their own dot cards for the grid is a further invitation for them to make different totals and to describe what they notice and to suggest what they might like to try next. This often results in cards being made with many more dots drawn than I would have anticipated or provided. Sometimes children become engrossed in who has a card that is *the same* as theirs. Often, children choose to count the total number of dots on their grid, involving them in counting to very high numbers with varying degrees of accuracy, enjoying writing the large totals and discussing who has made the largest number. Although the task of totalling the dots might become challenging, children often persist, finding ways of doing so and making new cards if necessary. They persist because they are invested in what is playing out; they are in control. Their confidence and playfulness are clear and other children want to join in to experience this. Here, control over the direction and focus of the task is shared, involving interaction between play (with and without an adult's presence) as well as input from a practitioner.

BALANCING TEACHING AND PLAY

Teaching should not be taken to imply a 'top down' or formal way of working. It is a broad term that covers the many different ways in which adults help young children learn. It includes their interactions with children during planned and child-initiated play and activities: communicating and modelling language; showing, explaining, demonstrating, exploring ideas; encouraging, questioning, recalling; providing a narrative for what they are doing; facilitating and setting challenges. It takes account of the equipment adults provide and the attention given to the physical environment, as well as the structure and routines of the day that establish expectations.

(OFSTED 2019a: 33; 2019b: 80)

I would consider having this definition displayed prominently in all Early Years settings, Nursery and Reception classrooms. I have found this to be an accurate and useful definition of the craft of teaching to discuss with senior school leaders or colleagues who may not have a background in Early Years education, particularly given the powerful body it comes from (the Office for Standards in Education or OFSTED). We can open up a discussion over this definition: what *teaching* can we observe over a morning, for example? Do these examples of teaching occur across all age groups? We can then ask them to tune into our young learners by spending short periods in our settings with a clear objective of what to observe, for example: writing down what children say and do within 15 minutes of block play. This can be used as a focus for discussions afterwards in relation to these three questions:

- What is there to see mathematically?
- How best can we understand what we see?
- How can we put these understandings to good use?

Figure 4.16

Figure 4.17

In OFSTED's definition of teaching, the role of quality interactions is recognised in all sorts of contexts, including playful ones. Adult-led teaching is indispensable for effective mathematics learning, but this should be unpressurised and playful, to avoid the development of maths anxiety (DEANS for Impact 2019). Frequency of experience in a wide range of contexts is important (WWC 2013) and continuous provision, in particular, offers the space for children to make sense of what they are being taught, to reinforce ideas and to do so independently. Moreover, through sensitive interaction in continuous provision, we can build young children's mathematical understanding. In the following vignette, the interaction of play and teaching is clear. Context draws children into the mathematics: it is the way into mathematics for all. Gradually, from time to time, we can strip away the context to reveal and connect with the abstract. The horse-jumping vignette that follows is a good example of this, with the children moving between activity and formal recording, relating them to each other.

VIGNETTE: HORSE JUMPING

Figure 4.18

A group of Reception children are outdoors. They are playing at taking part in a jumping competition on home-made hobby horses. Chloe and Reena are scoring; Reena is holding two half-coconuts to make a galloping-hoof noise. Ji and Eva are both on hobby horses, and Eva is wearing a riding hat.

Chloe: (*Holds up her two hands to show ten fingers*) Is it that much?

Reena: I'll show you on my fingers, OK?

Ji: Two minutes.

Reena: And it's a really fast two minutes. (*to Eva*) Off you go!

Eva starts the course and Ji waits for his turn. Someone cheers. Reena is commenting:

Reena: You got two faults Eva! Oh, and another two faults. And, and another two faults! Oh!

Chloe: You got six.

Reena: How many did she get? (*pulling disappointed face*) Oh, you only got four, Eva. So, can you write your score?

Chloe: You got four score.

Reena: So, you only got four that time. Off you go, off you go.

Eva goes to the scoreboard and writes. Chloe watches her, handing her a choice of pens. Eva writes '10 − 6 =' and pauses. Chloe points to it.

Chloe: (*following the sentence with her finger*) Four.

Figure 4.19

Eva writes '4', stands and smiles.

Reena: OK, Ji! And off you go! Ji, you've only got two minutes. Oh, you've only got two minutes Ji, go go go go go go go go!

Ji is going around the course on his hobby horse; Reena is banging the coconuts. Someone is calling, *Ji, Ji, Ji, Ji ...*

Figure 4.20

In this example, the children are running and jumping around outside, involved in self-chosen and self-directed play. The adult has tuned into what she has observed the children are currently interested in (playing horses and physical activity) and built some quite complex mathematics into this, subtracting scores from ten. It is play that has

Figure 4.21

been initiated by children, added to by an adult and further directed and managed by the children. It evolved over days as a collaboration between the adults and the children. In her interactions as time goes on, the adult invites children to think further by saying, for example: *You have thought hard about how to record that, but what does this say here?* or explained some conventions, for example: *When I write this, it is a quick way of writing 'take away'.* Research is clear that it is not straightforward for 4- and 5-year-olds to use formal symbolisation with understanding (Asmussen et al. 2018; Hughes 1986; Nunes et al. 2009; Thompson 2008).

This vignette is the result of sensitive interactions between intentional teaching episodes and well-planned continuous provision. One of the adults kept horses and the children had been asking questions. The practitioner then noticed children playing at 'being horses' by 'cantering' around the outside area and they asked her for some jumps. Group teaching and play built up over a couple of weeks with the mathematics developing from the adult introducing the children to jumping over and counting a small number of hurdles and recording this on a scoreboard, to numbered hurdles (one to ten and then above) being 'cleared' in order and to recording the number of jumps cleared as a final score. Finally, as seen in the vignette above, children counted 'faults' (jumps not cleared) to subtract from a starting total of ten. The mathematics was developed gradually and systematically, over a period of weeks, each time building on what had gone before. The practitioner introduced props and ideas and drew the children's attention to some mathematics which she related to a horse-jumping competition. She observed how the children played with these ideas, holding regular review sessions to hear what the children had to say about what had been happening. Importantly, she had a clear idea of the possible trajectory of learning for this area of mathematics, basing the initial prompt (how many jumps have you knocked down?) on some mathematics that is firmly within this age of child's experience (counting accurately within ten).

Studies have found that when teachers engaged in brief introductions to concepts, modelling what to do, children later took up these ideas in their play (Klibanoff et al. 2006). This is not the same as having a short adult-led session and following this with the same task set at a table. This is planning the environment to stimulate mathematics, which in turn feeds into our direct teaching episodes. One guaranteed way of incorporating our direct teaching of counting quantities into continuous provision is to make the inclusion of a scoreboard a regular feature, indoors and out: alongside dropping sponges into buckets, fishing treasure out of sand, bowling skittles, throwing beanbags at a target and so on. What is key is that the 'scores' refer to items that can be physically counted and amounts compared (beanbags in the hoop, pieces of treasure in the net, etc.) and not just abstract 'points'. Early number provision will include children seeing and using numerals written in many different forms; on display, finding them in books, on a dice or calculator. It is important that adults both physically model using numerals for a purpose as well as how to write them. Too often the mathematics recorded is detached from both the activity and the manipulatives. These need to be developed together, with adults supporting opportu-

Figure 4.22

nities for children to freely explore mark-making and mathematical graphics whilst they play (Carruthers and Worthington 2005; Worthington and van Oers 2016). The resulting written recordings (and for example, photographs taken of scoreboards) can be used to stimulate discussions after the action: *Can you tell us what happened here, Raoul, when you wrote this? Can we see who scored the most?* In this context, informal, playful mark-making can be used to support children's developing understanding of the role of mathematical symbols, linking numerals meaningfully to quantities.

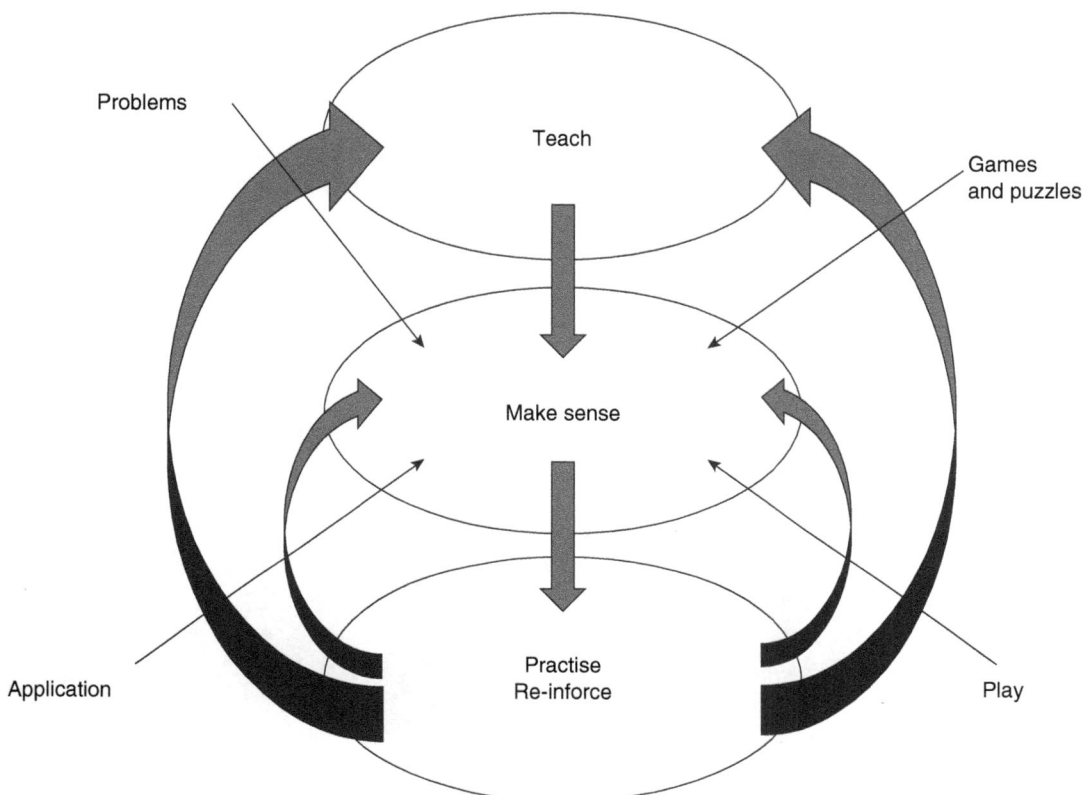

Figure 4.23

Children do not stop needing the space continuous provision and exploratory play provide to make sense at a certain age, or when they leave Early Years or Year 1: at all ages we need to provide the space for children to make sense of what they are being taught.

The diagram in Figure 4.23 considers the relationship between direct teaching, practice and sense making. Sense-making tasks might include any of the following:

- re-doing what has been taught without adult support;
- re-framing the taught task using different apparatus or different numbers;
- re-phrasing the teaching as a game or as a problem to solve.

We might call these opportunities to sense-make continuous provision, exploratory play or independent challenges and they are just as important right the way through Early Years and primary education. Importantly, they give us a valuable insight into what our children understand and can do independently. Here are three examples of sense-making opportunities I have used with 6- and 7-year-old children:

- modelling a mathematics game, leaving out these resources and encouraging children to play this independently and to teach others how to play;
- reading a mathematical picture book and leaving out the book with linked resources to explore, for example, *One Hundred Hungry Ants* (Pinczes and MacKain 1993) alongside 100 beadstrings, a tub of coloured pegs, 10×10 baseboards and big sheets of 1cm squared paper, to count out to 100 and recreate some of the ant arrangements;
- taking photographs of block structures or patterns made by children and displaying them as an invitation to design similar or different ones.

Next in this chapter on what we do as adults, I consider the role developmental learning trajectories can play in helping us tune into young children.

THE BIG IDEAS: THE 'WHAT' – LEARNING TRAJECTORIES

Figure 4.24

Behind the balancing of teaching and play lie two principles:

1. Having a clear sense of the interests, needs and disposition of the children for whom you are tailoring the teaching;
2. Having a clear understanding of the mathematical 'Big Idea' and a specific understanding you want to build.

(Early Math Collaborative, Erikson Institute 2014)

There is a very real need for confidence-boosting professional development for practitioners focused on the effective teaching of early mathematics. This is sadly not often readily or widely available. Knowledge of research-based developmental *learning trajectories* has been found to be important in effective teaching (Askew 2019; Clements and Sarama 2004, 2021; DEANS for Impact 2019; WWC 2013). What are these and how do we make use of them?

'Learning trajectories' are a learning construct and have their roots in previous theories of learning and teaching, such as Vygotskian *zones of development* (Vygotsky 1978), Deboys and Pitt's (1979) lines of mathematical development and Denvir and Brown's (1986) diagnostic description of number development for underachieving 7–9-year-olds. Learning trajectories have been developed from research into what children know and can do and outline typical, broad developmental paths for children's understanding towards a goal in specific mathematical areas. Whilst everyone's learning is much more like a spider's web than a straight line, and individual children move at different rates and take differing paths through these trajectories, broad lines of development can be observed. Rather than small 'steps' for every child to work through, these broad, hierarchical bands are for *practitioners* to get a good sense of the development of an area of mathematics. Importantly, they are not about content alone, but are built on *how* children come to understand. As such, they are *developmental* progressions. Take the example of oral counting: young children progress from saying some number words and responding to *one*, *two* or *lots*, through:

*   using counting 'strings' in play and everyday contexts;
*   reciting (chanting) numbers in order to five and above, recognising numerals of personal significance;
*   accurately 'tagging', saying one number for each item;
*   counting verbally beyond ten, beginning to spot repeating patterns;
*   recognising numerals from zero to ten, relating these to quantities, their size and order, and so on.

These are not progressive steps to be 'ticked off' as they are 'achieved', but broad paths of development, a landscape, to be taken into account when we work with our children, and all trajectories interrelate. Research is becoming clear that it is ineffective to simply teach young children what we want them to know and not pay attention to these trajectories, whereas it is highly effective to teach with these in mind (Clements and Sarama 2021; Siemon et al. 2019). Teaching with knowledge of a learning trajectory is both intentional and playful: it makes us experts in what we do and confident in the moment to follow children's interests. It helps us recognise interesting responses from children and the significance of these as well as knowing what to look out for. When we understand how children develop mathematical understanding by familiarising ourselves with these developmental paths, we are likely to be far more successful in:

- providing effective tasks and experiences;
- interacting sensitively and positively with children in their play;
- helping children reflect on their learning and deepen their understanding.

Clements and Sarama are prominent American researchers exploring learning trajectories at depth, producing many articles and books; they also have an extensive (free) website: **www.learningtrajectories.org**. Whilst we certainly do not need to log each child's movement through these trajectories, as this would be both unnecessary and extremely onerous, becoming familiar with both the general direction and the goal of learning in an area of mathematics we are intentionally teaching at that time will help us focus and adjust our inputs effectively for different groups of children. For example, thinking about the counting trajectory above, we might have collected a basket of leaves together. With some children we will be focusing our input on recognising one and two: *show me two leaves you like*; *show me two leaves that are big*; *show me two leaves that are the same*. With other children we will be focusing on them lining up a few leaves and 'tagging' each one accurately as they count them, or even taking a handful and them finding out who is holding the most. Thus, we deepen our knowledge of both the mathematics and of the children we work with. This is child-centred mathematics teaching, where the focus is on *learning* and fitting our teaching to what we observe. Moreover, we need to be aware that, at any point in time, for different children, some material we work with will be new, whereas some will be material that needs consolidating. We need to be mindful that we provide a curriculum that is *experiential,* as well as content we might expect them to 'know' in depth. For example, alongside the counting input above, we also make sure we talk about huge numbers in the thousands, people's ages, look at hundred-squares and have experience of very large collections.

    We can outline seven main Big Ideas in early mathematics. (The Early Math Collaborative has a slightly different list: **https://earlymath.erikson.edu/why-early-math-everyday-math/big-ideas-learning-early-mathematics/** and there are significant overlaps between all Big Ideas.)

The first four listed here are related to number:

- **Counting:** reciting the counting words correctly in sequence and using this to check each single item one by one to find the total number.

Figure 4.25

- **Cardinality:** establishing accurately how many items are in a group and understanding that the last number you say tells you how many there are.

Figure 4.26

- **Comparing:** knowing about the value of numbers in relation to each other.

Figure 4.27

- **Composition:** learning to recognise the small whole numbers that make up a larger whole number.

Figure 4.28

- **Measures:** making comparisons of length, weight (mass), time, capacity and volume and learning about units of measurement.

Figure 4.29

- **Shape and space:** composing and de-composing two- and three-dimensional shapes, understanding the effects of movement and combining shapes (spatial relationships) and developing the skills of spatial visualisation and orientation.

Figure 4.30

- **Pattern:** mathematics is based on an understanding of pattern and structure and developing an awareness of pattern and mathematical (algebraic) relationships. Pattern refers to any predictable sequence found in both geometric and numerical situations.

Figure 4.31

Developmental learning trajectories are a combination of child development research and instructional advice on how best to support this. They have a 'landscape' and a goal in mind, but they are not about 'teaching to' a goal – quite the opposite. Trajectories are interwoven and support each other. Our job as practitioners is to have our 'landscape of learning' in mind and weave our knowledge of *what* we teach together with *how* we teach it; if you like, the 'warp' of our woven curriculum, which would fall apart without the 'weft' running across it – the how – which I explore next.

THE BIG IDEAS: THE HOW – THE CHARACTERISTICS OF EFFECTIVE TEACHING AND LEARNING

Figure 4.32

- **playing and exploring** – children investigate and experience things, and 'have a go'
- **active learning** – children concentrate and keep on trying if they encounter difficulties, and enjoy achievements
- **creating and thinking critically** – children have and develop their own ideas, make links between ideas, and develop strategies for doing things.

(Department for Education 2021: 16)

These three Characteristics of Effective Teaching and Learning (CoETL) are an integral part of the statutory Early Years Foundation Stage in England. Both practitioners' actions and the environment they provide need to nurture these CoETLs. However, not only does their significance often come to an abrupt halt at the end of the Reception year, but the focus on them often does not extend to mathematics.

The CoETL are characteristics of effective *teaching* as well as learning. In what follows we take a look at the *teaching* we can do to develop these three characteristics in relation to learning mathematics.

Designing mathematical tasks with the three CoETL at their heart is a powerful way to place problem solving and reasoning at the heart of the mathematics we engage in with our young children. Creating and thinking critically is fundamental to mathematical thinking, but this does not occur without us nurturing all three characteristics alongside one another and none occurs by chance: our interactions must nurture these.

Figure 4.33

Some schools have used Table 4.1 as a basis for discussion with colleagues throughout the school to evaluate whole-school mathematics teaching, and to ensure significant attention continues to be paid to mathematical reasoning as the children move through the school. For example, are we allowing sufficient time mathematically for children to choose ways to do things? to make decisions about how to approach a task? To test their ideas?

Table 4.1 Table developed as part of an Early Years information pack for inspectors, relating the Characteristics of Effective Learning to mathematics (Ofsted 2013)

Characteristics of Effective Teaching and Learning	
Playing and exploring – engagement	**Finding out and exploring**
	Playing with what they know
	Being willing to 'have a go'
Active learning – motivation	**Being involved and concentrating**
	Keeping trying
	Enjoying achieving what they set out to do
Creating and thinking critically – thinking	**Having their own ideas** • Thinking of ideas • Finding ways to solve problems • Finding new ways to do things
	Making links • Making links and noticing patterns in their experience • Making predictions • Testing their ideas • Developing ideas of grouping, sequences, cause and effect
	Choosing ways to do things • Planning, making decisions about how to approach a task, solve a problem and reach a goal • Checking how well their activities are going • Changing strategy as needed • Reviewing how well the approach worked

Table 4.2 provides examples of what practitioners might say and do to extend shape, space and pattern play to encompass and nurture the CoETL.

I hope you find these two tables helpful in reconsidering the integration of the Characteristics of Effective Teaching and Learning with your mathematics provision. The next section looks more closely at adult–child interactions.

Table 4.2 Examples of what practitioners might do and say to integrate CoETL into spatial play

Characteristics of Effective Learning	Examples of what practitioners might do or say
Playing and Exploring: teaching suggestions	
• Children investigate and experience things, and 'have a go'.	
Finding out and exploring:	**Encouraging engagement by playing with resources yourself:** *What if … we use these blocks to make an interesting pattern?* *I wonder … if we can make all these shapes fit together?* *What shall we try (now)?* *Come and have a play with this.*
Playing with what they know:	*You covered the whole table using all the yellow hexagons yesterday, what do you think you might do today?* *What did you think about trying?*
Being willing to have a go:	*This might take a while! Shall we start?* *What about something taller, longer, larger …?* *This is a jigsaw with lots of pieces; shall we try it?*
	Model making ludicrous suggestions, deliberate mistakes: *I think this is the only pattern that we can make!* *There isn't any piece that will fit there!* *I am going to make my tower touch the ceiling!*
	Celebrating a range of solutions for problems: *What shall we try and make?* *How shall we …?* *Why not give it a try and see what happens?* *Two patterns that repeat! How is that one different to/the same as, yours?*
Active Learning: teaching suggestions	
• Children concentrate and keep on trying if they encounter difficulties and enjoy achievements.	
Being involved and concentrating:	**Celebrate perseverance:** *You are looking carefully at where to place each shape so that it fits/matches.*
Keeping trying:	**Celebrate alternative strategies:** *What will you try? And what about you, Evan?* *So, you cut that out and it's not a square, what will you try next to make sure it's square this time?* *You have spotted that pattern doesn't 'work'; how might you make it work?* *Why not give it a try and see what happens?* *Two patterns that repeat! How is this one the same as/different to yours?*

Characteristics of Effective Learning	Examples of what practitioners might do or say
Active Learning: teaching suggestions	
Enjoying achieving what they set out to do:	**Encourage reflection:** *Was that what/did that turn out as you expected?* *That was quite a struggle - well done for sticking at it.* *Did you change your mind? That's often a good idea. What did you end up doing?*
Creating and Thinking Critically: teaching suggestions	
• Children have and develop their own ideas, make links between ideas, and develop strategies for doing things. ***Having their own ideas:*** *thinking of ideas,* *finding ways to solve problems,* *finding new ways to do things*	**Giving children the responsibility to solve problems:** *What might we do? What will we need? What shall we try (now)?* *How might we fill that gap?* *What might your maze/castle/pattern look like, do you think?*
Making links: *making links and noticing patterns in their experience,* *making predictions, testing their ideas, developing ideas of grouping, sequences, cause and effect*	*Isn't that a little bit like that smaller model you made yesterday?* *That pattern reminds me of ... counting in twos!* *How is this one similar (different)?* *What do you think might happen if ...?* *What if you ...?* *I wonder, if you ... then what might it look like?* **Reminding them of the problem:** *Is this going to work, do you think?* *What are you trying to do?* *Is there a better way of covering this space?* *What about if we turn that around/upside down/replace that piece?*
Choosing ways to do things: *planning, making decisions about how to approach a task, solve a problem and reach a goal,* *checking how well their activities are going,* *changing strategy as needed,* *reviewing how well the approach worked*	**Reflecting on how it has gone:** *Can you say what you like about the model?* *What might you try next? Why?* *How could we have built that doorway in a different way, so that ...?* *I thought that might work; what did you think at first? ... later on?*

CONVERSATIONS

Here we explore the nature of our mathematical interactions with children; interactions that include conversations, modelling and developing moments of *sustained shared thinking* (Siraj-Blatchford et al. 2002), *re-proposal* (Malaguzzi 1993), metacognition and reflection. To create a situation where children have the space to say interesting things, we have to firstly provide something mathematical we think might fascinate them by tuning into their interests and ideas, and secondly, we have to be quiet and wait for the child to speak, and then be quiet some more in order for other children to join in. Cheeseman worked with highly

Figure 4.34

effective mathematics teachers of 5–7-year-old children and came to the – somewhat unsurprising – conclusion that adults should endeavour to:

plan time for close mathematical conversations with children, expect thinking of children, including conjecturing, reasoning, justifying, and consider tasks and their potential to engage and extend children's thinking.

(Cheeseman 2018: 22)

Our mathematics tasks need to stimulate what they refer to as *thinking conversations*, with the adult playing a key role in eliciting young children's mathematical thinking through sensitive repetition and some carefully chosen questions. *Thinking conversations* involve us being curious about our children's thinking rather than questioning children to pursue an answer. They are what we recognise in Early Years as *sustained shared thinking* (Siraj-Blatchford et al. 2002). Sustained shared thinking involves:

- the practitioner having an awareness of, and responding to, a child's understanding and capabilities;
- the adult *lifting the level of thinking*;
- the practitioner and child together developing an idea or skill.

Figure 4.35

The 'Researching Effective Pedagogy in the Early Years' research pinpoints sustained shared thinking as characterising the most effective pre-school settings (Siraj-Blatchford et al. 2002). Such conversations do not have to be long or take up a lot of time; even brief conversations can open up children's thinking. These might involve practitioners doing any of the following, some of which you will recognise from the vignettes throughout this book:

It is worth reflecting on which of these we make use of in our mathematical interactions with our children.

- *inviting children to elaborate: I really want to know more about this pattern* … (see the vignettes 'Dotty cards' and 'Re-proposing' in this chapter);
- *re-capping: So you think that this will fall over if I add another brick?* (see the vignette 'Lemonade factory' in Chapter 2);
- *offering your own experience: I like to jot things down when I am trying to remember something* (see the vignette 'Horse jumping' in this chapter);
- *clarifying ideas: Right, Darren, so you think that this stone is heavier than this one, but not as heavy as this one; do I have that right?* (see the vignette 'Counting over 100' in Chapter 2);
- *suggesting: You might like to try doing it this way* (see the vignette 'Observing a 3-year-old with blocks' in Chapter 2);

- *reminding: Don't forget that you said that this stone is heavier than this one* (see the vignette 'Racing' in Chapter 2);
- *explaining: When I roll up this sheet of paper evenly like this, it makes a cylinder* (see the vignette 'Horse jumping' in this chapter);
- *using encouragement to further thinking: You have really thought hard about how many we need, but why is it that Raj is saying a different number?* (see the vignette 'Horse jumping' in this chapter);
- *speculating: Hmmm. Do you think this cup will hold more beads or more corks?* (see the vignette 'Dotty cards' in this chapter);
- *offering an alternative viewpoint: Maybe Goldilocks wasn't naughty when she broke the chair;*
- *reciprocating: Thank goodness you put each counter in that box as you counted them – I keep losing count!;*
- *asking open questions: What do you notice? What do you wonder? How did you …? Why do you think …? What might happen next?;*
- *modelling thinking: I have to think hard about what to do this afternoon. I have a lot of things to do. I need to take my dog to the vet's, take my library books back to the library and buy some food for dinner tonight. But it will be hard to fit in all those things, so …;*
- *reflecting: I wonder if …; I wonder why?*

Figure 4.36

With younger or less-experienced children it can be easy to underestimate the value of modelling our thoughts. Commentating on play is a sensitive way to 'drop in' new vocabulary. Interesting longitudinal research undertaken with mothers and their 2-year-old children (Ribeiro et al. 2020) found that higher levels of verbal support during mother–child interactions when solving jigsaw-type puzzles together predicted the children's wider mathematical achievement when they were 7 years old. Mothers were asked to support their children using spatial language such as straight, round, small, etc., alongside gesture. An example of what this research team referred to as *high quality spatial interaction* for the puzzle task involved both statements and gestures such as:

> *Try placing it next to that one, now it is upside-down, you need to turn it around, that's it, yes, great!; Try above the tractor, it's upside-down, try the other way.*

(Ribeiro et al. 2020: 290)

Talking alongside young learners whilst they build, pattern-make and block play, introducing mathematical and descriptive language as well as modelling our thinking, is an important role for adults to adopt. Unobtrusive responses could include: *Remember these two are the same size as this one*; *So if that one doesn't fit*; *what about trying this shape? How about we turn that triangle around?* Once we have opened up a conversation the challenge then is to not close it down, but rather to sustain it into a to-and-fro interaction, a conversation that is cumulative and builds. Invitations to a child to elaborate, or re-capping what you think you have heard them say, as well as reflecting their thinking back to them (vignette 'Re-proposal'; see below) are powerful here.

Mathematical learning requires children to make connections and this involves an element of reflection over and above activity. *Metacognition*, or thinking about our thinking, was long thought to be out of reach for very young children. However, Robson (2010) and Williams (2014) videoed episodes of self-initiated play activities as prompts for semi-structured 'reflective dialogues' with 4- and 5-year-olds. With Gifford (2004) and others (Griffiths 2011; Tanner and Jones 2007), they found evidence of metacognitive behaviours in these young children, challenging the view that such abilities are later in developing. Watching video clips of their play provides an opportunity for children to focus, reflect and discuss themselves thinking. The use of a still or moving image taken of a few children and shown to a wider group can be an effective way to prompt reflection on, and stimulate further thinking about, mathematics (Williams 2014). Such images act as prompts for us to find out what the children might know and understand as well as providing time for children themselves to reconstruct their knowing and push it forwards.

One powerful example of developing Early Years reflective practice is the strategy of *re-proposal* used in the Italian pre-schools of Reggio Emilia (Malaguzzi 1993). Re-proposal is the final part of the three Reggio Emilia principles of *observing* children, *documenting* their learning and *re-proposing* their thinking. It is the act of re-playing or reflecting learners' thinking back to them. In re-proposal, adult observers choose and note accurately a short piece of overheard child's speech, later reading this back to the child for them to enlarge on and investigate further. Proponents of this strategy are clear that this should be done with *no* adult interpretation or additional comment, thus leaving the space for the learner to think about thinking (Abbott and Nutbrown 2001). In the vignette that follows I use a video clip to 're-propose' for these young children to focus on their mathematical thinking.

Teach me about that and *Help us learn about that* are phrases I have developed, alongside others listed on pages 151–152, to invite children to say more – to explain – in both one-to-one and group situations. I have found they work as long as I then keep quiet and listen.

VIGNETTE: RE-PROPOSAL

Figure 4.37

Figure 4.38

Rachel and Sabina are both 5 years old. They have been timing themselves racing around outside with a digital stopwatch. The adult shows them an image she has taken that morning of them involved in the play.

Adult: Can you teach me about that?

Rachel: (*immediately*) The time, the time and when it stops you know what the time is how much you did it, so like, if you got the smaller number, that means you're faster and if you're, and if there's a bigger number that means you haven't beat Max, that means it's a bigger number.

The child is often keen to expand on their original statement partly because it puts them in the position of being expert and knowledgeable. It also makes clear that I value what they are thinking. To encourage children to listen to what other children have to say, I can follow this by saying, *Who heard what Rachel said well enough, that they can say it in their own words?* Importantly, this is not said as a way of 'picking on' those who are not listening, but to encourage re-phrasing. If this is an unfamiliar request, I usually have to ask the first speaker to repeat what they have said, so we can all hear, sometimes more than once. I always check in with the speaker if they agree that is what they actually said or meant. Such strategies take time to embed but are an important way of encouraging listening conversations with our young learners. Even young children become used to being asked and to answer these searching questions when they are asked often and sensitively enough.

In the re-proposal vignette on page 155, the adult showed the children a video clip of the play to prompt some reflection and we can see Rachel striving to explain and generalise the complex mathematics of speed. The activity is not the end result, but is followed by an opportunity to reflect, by looking again at what children have done, drawn or written, by re-proposing. Providing opportunities such as these gives even the youngest child, who may not yet be as articulate as Rachel above, the chance to pause and to re-look at what they have been involved in. Hendy and Toon put this well, saying that:

Real life moves too fast.

(Hendy and Toon 2001: 63)

Reflection does not have to take up an enormous amount of time, neither is it necessary after every period of activity, but regularly used it is valuable.

Conversations we hold with our young children make important contributions to their growing awareness of the mathematics they know and can do. Providing children with regular, brief opportunities to reflect on what they have been involved in and practitioners regularly discussing the children's play with them during group times might not only help children deepen their learning, but also make a contribution to their positive self-image in relation to mathematics, by emphasising that what they say and do is important and is being listened to.

TO SUMMARISE

As adults in Early Years settings and classrooms, both the mathematical provision we decide upon and the culture of our settings (how we respond to children) are crucial in developing young children's mathematics. Every one of our young children can learn sophisticated mathematics in a supportive yet challenging environment with an emphasis on low-threshold, inclusive tasks which we know how to develop and extend. This chapter has been about our role as adults in our mathematical interactions with young children. Both appraising ourselves of mathematical learning trajectories and being observant of and responsive to what children are interested in and capable of, are important. Interweaving longer observations with shorter, more instant 'noticings' is informative. Marion Bird neatly summarises how she sees the role of the adult in the following quotation:

I have found it useful to focus my attention on six main areas: the opening of an activity; planning sessions; seeking and using children's ideas; aspects of the teacher's authority; putting children at ease; and involving children in complex and sometimes conflicting ideas.

(Bird 1991: 123)

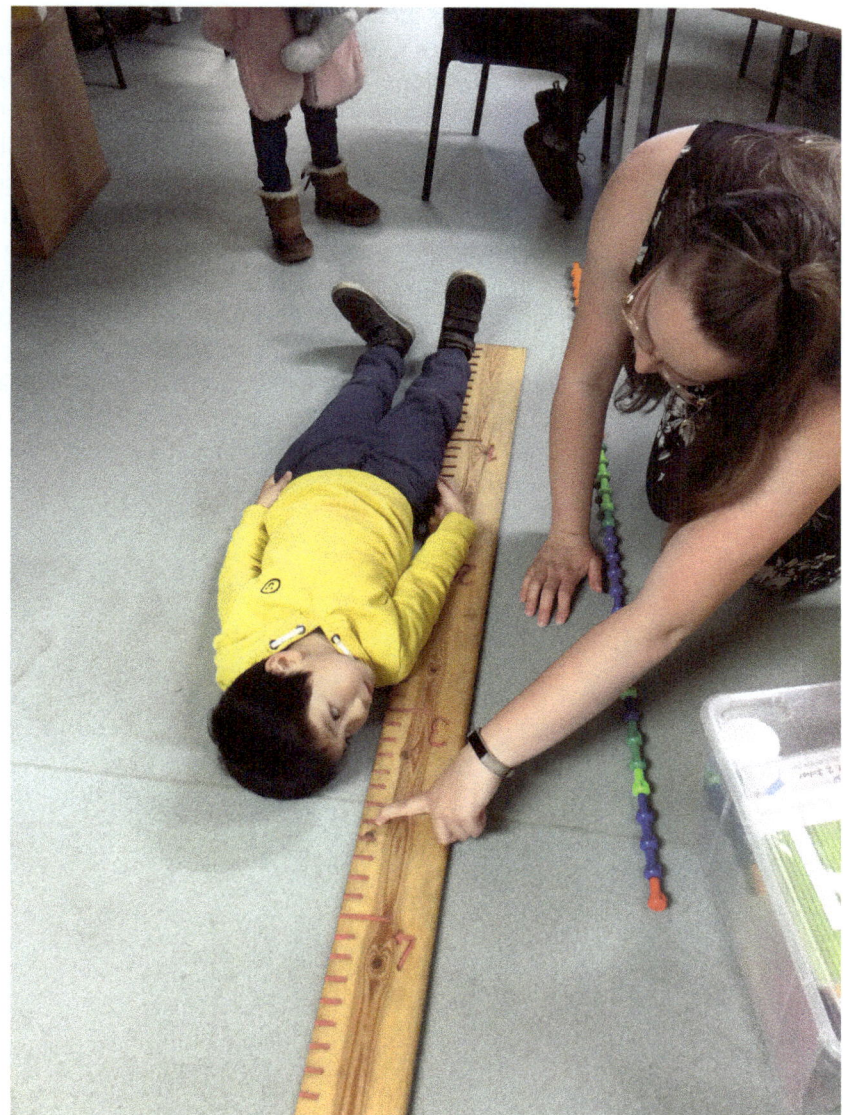

Figure 4.39

This is an important list from Marion Bird:

- preparing how to begin;
- seeking and using children's ideas;
- putting children at ease;
- involving children in complex and sometimes conflicting ideas.

The mathematical environment, as well as being *what* we provide, is about *how* we work within that, the relationships we set up and the culture we establish around mathematics. We balance our mathematics teaching episodes between those that are more adult-directed and those where we allow children to direct what plays out. Within mathematical episodes, we move between explaining, demonstrating and modelling, and encouraging children to explore their own ideas. We provide space and time later for children to think about what they are learning. Valuing all young children's ideas gives them ownership and increases their engagement and perseverance, as well as building on what they understand and helping them form positive mathematical identities. Enjoying your own mathematical explorations with children will model constructively how it feels to play around with some mathematics. It is important we find tasks *we* like to take part in and questions *we* are interested in exploring. And lastly, we make sure our mathematical conversations with children, however brief, respect each child's evolving mathematical ideas. By paying attention to a child as an individual we might also make a contribution to their positive self-image in relation to mathematics.

5

SOME CONCLUDING REMARKS ON BEING MATHEMATICALLY PLAYFUL

THE SUBSTANCE OF TEACHING

This book is a book about pedagogy and curriculum – the substance of teaching. Its key elements are tuning in to children, knowing your stuff and becoming playful around mathematics. It is about roller-skating, as well as tidying your room. This book is aimed primarily at those of us who work with 3- to 7-year-olds, children near the beginning of their mathematical journey, and below the formal school starting age in many countries. I believe much of what I have written applies to older, even much older, learners. As I have written this book, I have become increasingly aware of the downward pressure on our youngest children, particularly those between the ages of 4 and 7, of approaches that may be suitable, on occasions, for much older children. This is concerning, as this pressure fails to take into account that when we work in a developmentally appropriate way with the children in front of us, effective learning takes place. We have reached the unfortunate position in the twenty-first century, in England in particular, of placing excessive emphasis on requirements for children to achieve certain targets at fixed points in time. Whilst it is useful to have some benchmarks to help monitor whether (and which) children are learning appropriately for their age and stage, in particular, for more disadvantaged children, the problems arise in our current 'high-stakes' environment when these are used for school accountability purposes. Here the usefulness of a target is distorted as pressure is put on as many children as possible to achieve an unrealistic (or highly unlikely) objective. Alongside this, we have a general lack of understanding within the population of effective Early Years practice and recognition of the substantial knowledge and skills that young children bring to school (Asmussen et al. 2018; Goswami 2015; Nunes et al. 2009). It is difficult, as the sole Early Years practitioner in a setting or school, to withstand pressures which appear to run counter to what we observe and what we know and feel is right. To effectively withstand pressure, we have to educate ourselves and our colleagues about what we do and know. We can all strive to be colleagues who share, support, enrich and propose effective mathematical ways forward based around the children in our care.

We all need to:

- learn about early mathematics and become informed in terms of what research tells us: there are many references included in this book which I hope will help. Becoming familiar with a learning trajectory approach (as opposed to a small-steps or 'progression' model), as articulated by Doug Clements and Julie Sarama (2021), is key. Clements and Sarama conceptualise learning trajectories as a powerful combination of developmental progressions and instructional activities to support children as they move through these developmental progressions. This is quite different to breaking down an end goal into substeps of skills based on an adult perspective;
- articulate clearly what we know: exploring and talking about developmental learning trajectories with colleagues is a good way of doing this;
- make connections with other Early Years practitioners as well as colleagues in other stages, sharing our experiences and observations, in person and through social media.

First and foremost, we must have the child at the centre of what we decide to do. In order that children become resilient and confident, as well as competent, mathematicians I believe it is essential that they learn to enjoy and feel in control of *their* mathematics; that it is something *they*, not just we, do. This requires that *we* are well prepared, as opposed to us having a well-prepared lesson. It requires:

> *having good subject knowledge, knowing misconceptions, knowing your pupils and being open to situations that will challenge everyone in your classroom including you.*
>
> (Griffin 2019: 17)

This preparation will afford us the confidence and the enjoyment of going 'off-piste' when required, and to genuinely respond to our children and what they are doing and saying in that moment.

Developing a playful attitude to our mathematics interactions, regardless of whether it is the adult or the child who initiates these, will nurture a child's positive attitude to a subject commonly seen as difficult and dull. Classrooms and Early Years settings are emotional spaces and mathematics provokes a lot of negative emotion. It is important we all tackle this head on. It is not enough to say, as I have heard too often, *when they are successful, they will learn to enjoy it*, as by then it is far too late for many of our children. Mathematics anxiety occurs in a wide context and both emotions and cognition affect learning. Rather than simply teaching to inform, we should strive also to teach to inspire. What drives learning is curiosity, imagination, creativity and motivation. Given the opportunity, our young children can and do inspire us mathematically. We need to remember, to paraphrase the respected mathematics educator Marilyn Burns, 'We are teaching students, not the curriculum' (Listening to Learn 2021).

Historically and internationally, there have been two main issues affecting Early Years mathematics:

> *one is the inadequate preparation of Early Years educators and another is a widespread, and mistaken, assumption that young children are not interested in mathematics, nor are they capable of engaging in any kind of abstract mathematics.*

> (Moss et al. 2015: 2)

I hope this book helps convince you of the latter and contributes towards our preparedness for teaching effective early mathematics. I truly believe that having conversations with our colleagues

Figure 5.1

Figure 5.2

about what we notice our children doing mathematically, combined with the curiosity to find out where what we observe comes from and might go to, is the way we move forwards as a sector and withstand inappropriate external pressure. We have to become informed, articulate and brave. A number of helpful websites dedicated to Early Years mathematics are listed at the end of this chapter under 'Resources'; along with the References section these will point you to further reading, should you hopefully become as fascinated by early maths as I have become.

REACHING OUT FROM EARLY YEARS AND KEY STAGE 1

Naturally, the teaching of 3- to 7-year-olds does not exist in isolation. I have deliberately chosen to cross from Early Years Foundation Stage into Key Stage 1 because I know that effective mathematical practice is seamless as the children move between settings and classes, each building on the last (Fisher 2020). Also important is us building on what has happened before and what is happening outside school. This means making time to talk with colleagues alongside, 'behind' and 'in front' of us and creating time to connect with and involve families. I hope this book helps in these discussions.

It sometimes helps to have some questions to focus discussions with colleagues. Here are some I have collected that I have found helpful in the past:

- In what ways does this maths task encourage … children's maths chat … collaboration … reasoning?
- What do we have outdoors/indoors to develop this particular Big Idea of mathematics?
- What and where are our mathematical sense-making opportunities?
- What space is there for children's exploratory mathematical play?
- What reference do we make mathematically to the Characteristics of Effective Teaching and Learning?
- Do we (all) use continuous provision? Mathematically? Is it used to consolidate? To introduce something new? To further extend or enrich an idea? How might this develop over 2 weeks? How does it connect with and support my direct teaching? How does it connect with and support with children's own mathematical interests?
- When and why might I begin with a problem, or an open situation, for children to explore? What mathematical possibilities do I anticipate flowing from this?
- How do I assess what is playing out here?

Finally, it can be disquieting to set down what you believe in a book, something set in stone to be quoted – what if I change my mind? Most of what I write here has been forming over the last 40 years as an educator. As I trawled through all my journals and notes kept over this long period, I could spot threads of what I later researched in more depth. An important part of our development as educators is to continue to talk, listen, try out ideas and to change our minds, and I reserve the right to do so. Reflective practice is a large part of who I am. In writing this book I hold on to something John Mason, mathematics educator, is recognised as saying: *Everything I write here is a conjecture.* Be sure to check it out with your own experience. This applies to formal, published research as well as to anything else. Nothing should become something that cannot be questioned.

With our eyes on the big picture of knowing what we are doing, why we are doing it and how to do it well, we can, as a Reception teacher said to me recently, *enjoy the small moments of maths with our children* (thank you, Leonne Cummins).

Sam (4 years 9 months):

You know, I counted to a thousand once. I could count to a million, but I've never tried.

I end this book with the two questions I usually ask at the end of any professional development session:

What do I make of what I have read here?

What will be my next steps … as a professional … as a colleague?

USEFUL RESOURCES

WEBSITES

All of these are free to access, although some may require you to register.

UK

https://earlymaths.org
The Early Childhood Mathematics Group (ECMG) is a UK-based group of Early Years mathematics enthusiasts and experts, teachers, practitioners, researchers and professors who work together to promote early childhood mathematics. They have a number of guidance documents and a range of support for working mathematically with children between the ages of birth and 7 years.

https://nrich.maths.org
NRICH is a collaboration between the Faculties of Mathematics and Education at the University of Cambridge which focuses on problem solving and on creating opportunities for students to learn mathematics through exploration and discussion. It has a dedicated section for those who work with 3–5-year-olds.

https://creativestarlearning.co.uk
Creative Star Learning was established in 2007 by Juliet Robertson to provide *support, training, advice* and *resources* on almost all aspects of outdoor learning and play, hence the STAR in the company name.

www.mathsthroughstories.org
This is an international research-based initiative which explores aspects of integrating story picture books and imaginative story writing in mathematics teaching. Their mission is to *make mathematics teaching more accessible and enjoyable* for learners everywhere through the power of story telling and children's imagination by sharing practical ideas for teachers and parents on how to embrace this approach.

USA

www.learningtrajectories.org
The creators of Learning and Teaching with Learning Trajectories [LT]^2 are Drs Sarama and Clements, who are world leaders in the learning and teaching of early mathematics. This comprehensive website has developmental learning trajectories for each area of mathematics, from birth to 8 years of age.

https://dreme.stanford.edu
The DREME (Development and Research in Early Math Education) network is based at Stanford University and was created in 2014 to advance the field of early mathematics research. It is focused on mathematical development from birth to age 8 years of age, with sections for families as well as teachers and practitioners.

https://earlymath.erikson.edu
The Erikson Institute's background is in child development. In 2007 it launched the Early Math Collaborative to increase the quality of early maths education in three key ways: professional development, conducting research and information on foundational mathematics. There is a free newsletter and the site brims with ideas, videos and CPD opportunities.

REFERENCES

Abbott, L., and Nutbrown, C. (eds.) (2001). *Experiencing Reggio Emilia: Implications for Pre-School Provision*. Buckingham: Open University Press.

Anthony, G., and Walshaw, M. (2007). *Effective Pedagogy in Mathematics/Pangaram: Best Evidence Synthesis Iteration [BES]*. Wellington, New Zealand: New Zealand Ministry of Education.

Askew, M. (2008). Social identities as learners and teachers of mathematics: The situated nature of roles and relations in mathematics classrooms. In A. Watson and P. Winbourne (eds.) *New Directions for Situated Cognition in Mathematics Education*. New York: Springer.

Askew, M. (2019). Epilogue. In D. Siemon, T. Barkatsas and R. Seah (eds.) *Researching and Using Progressions (Trajectories) in Mathematics Education*. Boston: Brill Sense.

Askew, M., Hodgen, J., Hossain, S., and Bretscher, N. (2010). *Values and Variables: Mathematics Education in High Performing Countries*. London: Nuffield Foundation.

Asmussen, K., Law, J., Charlton, J., Acqua, D., Brims, L., Pote, I., and McBride, T. (2018). *Key Competencies in Early Cognitive Development: Things, People, Numbers and Words*. London: Early Intervention Foundation.

Back, J. M. (2004). *Mathematical Talk in Primary Classrooms: Forms of Life and Language Games*. PhD thesis, King's College London.

Bird, M. H. (1991). *Mathematics for Young Children: An Active Thinking Approach*. London: Routledge.

Boaler, J. (2013). Ability and mathematics: The mindset revolution that is reshaping education. *Forum* 55(1).

Borthwick, A., and Cross, A. (2018). *Reasons to Reason in Primary Maths and Science*. London: SAGE.

Borthwick, A., Gifford, S. and Thouless, H. (2021). *The Power of Pattern: Patterning in the Early Years*. Derby: ATM.

Brown, T. (1996). Play and number. In: R. Merttens (ed.) *Teaching Numeracy: Maths in the Primary Classroom*. Leamington Spa: Scholastic.

Carruthers, E., and Worthington, M. (2005). Making sense of mathematical graphics: The development of understanding abstract symbolism. *European Early Childhood Education Research Journal* 13(1): 57–79.

Casey, B., Andrews, N., Schindler, H., Kirsch, J., Samper, A., and Copley, J. (2008). The development of spatial skills through interventions involving block building activities. *Cognition and Instruction* 26: 269–309.

Cheeseman, J. (2018). Creating a learning environment that encourages mathematical thinking. In M. Barnes, M. Gindidis and S. Phillipson (eds.) *Evidence-Based Learning and Teaching*. New York: Routledge; pp. 9–24.

Cheeseman, J., and McDonough, A. (2016). Fostering mathematical curiosity. In S. Dockett and A. MacDonald (eds.) *Just Do Good Research: Commentary on the Work and Influence of Bob Perry*. Albury, Australia: Peridot Education; pp. 142–151.

Cheng, Y., and Mix, K. S. (2014). Spatial training improves children's mathematics ability. *Journal of Cognition and Development* 15(1): 2–11.

Clements, D. H., and Sarama, J. (2004). Learning trajectories in mathematics education. *Mathematical Thinking and Learning* 6(2) 81–89. Available at: https://www.researchgate.net/publication/233300911_Learning_Trajectories_in_Mathematics_Education

Clements, D. H., and Sarama, J. (2021). *Learning and Teaching Early Math: The Learning Trajectories Approach*, 3rd ed. New York: Routledge.

Cochran, K. F., De Ruiter, J. A., and King, R. A. (1993). Pedagogical content knowing: An integrative model for teacher preparation. *Journal of Teacher Education* 44: 263–272.

Davenall, J. (2016). *Young Children's Mathematical Recording*. Cambridge: NRICH. Available at: **https://nrich.maths.org/12384**

Davenall, J., and Gifford, S. (2015). *Developing Number Through Tidying Up*. Cambridge: NRICH. Available at: **https://nrich.maths.org/11528**

Davis, G., and Pepper, K. (1992). Mathematical problem solving by pre-school children. *Educational Studies in Mathematics* 23: 397–415.

DEANS for Impact. (2019). *The Science of Early Learning: How Young Children Develop Agency, Numeracy and Literacy*. Austin, TX: Deans for Impact.

Deboys, M., and Pitt, E. (1979). *Lines of Development in Primary Mathematics*. Oxford: OUP.

Denvir, B., and Brown, M. (1986). Understanding of number concepts in low attaining 7–9-year-olds (part 1): The teaching studies. *Educational Studies in Mathematics* 17(2): 143–164.

Department for Education. (2007). *Gender and Education: The Evidence on Pupils in England*. Available at: **https://dera.ioe.ac.uk/6616/8/rtp01-07_Redacted.pdf**

Department for Education. (2021). *Statutory Framework for the Early Years Foundation Stage: Setting the Standards for Learning, Development and Care*. Available at: **https://assets.publishing.service.gov.uk/government/uploads/system/uploads/attachment_data/file/974907/EYFS_framework_-_March_2021.pdf**

Donaldson, M. (1978). *Children's Minds*. London: Fontana.

Dowker, A. (2021). Home numeracy and pre-school children's mathematical development: Expanding home numeracy models to include parental attitudes and emotions. *Frontiers in Education* 6 (February): 1–9.

Downton, A., MacDonald, A., Cheeseman, J., Russo, J., and McChesney, J. (2020). Mathematics learning and education from birth to eight years. In: J. Way, C. Attard, J. Anderson, J. Bobis, H. McMaster and K. Cartwright (eds.) *Research in Mathematics Education in Australasia 2016–2019*. Singapore: Springer.

Early Math Collaborative, Erikson Institute. (2014). *Big Ideas of Early Mathematics: What Teachers of Young Children Need to Know*. Boston: Pearson

Education, Audiovisual and Culture Executive Agency. (2018). *Compulsory Education in Europe 2018/19*. Luxembourg: EACEA, Education and Youth Policy Analysis. Available at: **https://eacea.ec.europa.eu/national-policies/eurydice/sites/eurydice/files/compulsory_education_2018_19.pdf**

Education Scotland. (2020). *Realising the Ambition: Being Me. National Practice Guidance for Early Years in Scotland*. Available at: **https://education.gov.scot/media/3bjpr3wa/realisingtheambition.pdf**

Fisher, J. (2020). *Moving on to Key Stage 1: Improving Transition into Primary School*. Oxford: OUP.

Fosnot Twomey, C., and Dolk, M. (2001). *Young Mathematicians at Work: Constructing Number Sense, Addition and Subtraction*. Porstmouth, NH: Heinemann.

Garvey, C. (1977). *Play*. Cambridge, MA: Harvard University Press.

Gifford, S. (2004). A new mathematics pedagogy for the early years: In search of principles for practice. *International Journal of Early Years Education* 12(2): 99–115.

Gifford, S. (2006). *Teaching Mathematics 3–5: Developing and Learning in the Foundation Stage*. Maidenhead: Oxford University Press.

Gifford, S. (2015). *Early Years Mathematics: How to Create a Nation of Mathematics Lovers*. Available at: **https://nrich.maths.org/11441**

Griffin, P. (1989). Teaching takes place in time, learning takes place over time. *Mathematics Teaching* 126: 12–13.

Griffin, D. (2019). Teaching takes place in time, learning takes place over time: A response. *Mathematics Teaching* 268: 15–17.

Gripton, C., and Pawluch, D. (2021). Counting collections in the early years. *Mathematics Teaching* 275: 6–10.

Gopnick, A. (2015). *The Philosophical Baby: What Children's Minds Tell Us About Truth, Love, and the Meaning of Life*. Carnegie Science talk 1 June 2015. Available at: **www.youtube.com/watch?v=lUl0EG3hDVU**

Goswami, U. (2015). *Children's Cognitive Development and Learning. CPRT Research Survey 3*. York: Cambridge Primary Review Trust.

Gregg, S. (2020). *Pattern Blocks: A Tool for Mathematics Education*. Derby: Association of Teachers of Mathematics.

Griffiths, R. (2010). Money and shops, role play and real life. *Mathematics Teaching* 174: 20–22.

Griffiths, R. (2011). Exploring children's interest in seeing themselves on video: Metacognition and didactics in mathematics using 'Photobooth'. *Informal Proceedings of the British Society of Research into Learning Mathematics* 31(1): 61–66.

Griffiths, R., Back, J., and Gifford, S. (2016). *Making Numbers: Using Manipulatives to Teach Arithmetic*. Oxford: OUP.

Gura, P. (ed.) (1992). *Exploring Learning: Young Children and Blockplay*. London: Paul Chapman.

Haylock, D., and Cockburn, A. D. (2017). *Understanding Mathematics for Young Children*. Oxford: Blackwells.

Hendy, L., and Toon, L. (2001). *Supporting Drama and Imaginative Play in the Early Years*. In V. Hurst and J. Joseph (eds.) *Supporting Early Learning Series..* Buckingham: Open University Press.

Hogden, J., Pepper, D., Sturman, L., and Ruddock, G. (2010). *Is the UK an Outlier? An International Comparison of Upper Secondary Mathematics Education*. London: Nuffield Foundation. Available at: **www.nuffieldfoundation.org/wp-content/uploads/2019/12/Is-the-UK-an-Outlier_Nuffield-Foundation_v_FINAL.pdf**

Hough, S., and Gough, S. (2007). Realistic mathematics education. *Mathematics Teaching* 203: 34–38.

Hughes, M. (1986). *Children and Number: Difficulties in Learning Mathematics*. Oxford: Basil Blackwell.

Johnson, N. C., Turrou, A. C., McMillan, B. G., Raygoza, M. C., and Franke, M. L. (2019). 'Can you help me count these pennies?': Surfacing pre-schoolers' understandings of counting. *Mathematical Thinking and Learning* 21(4): 237–264. Available at: **https://doi.org/10.1080/10986065.2019.1588206**

Joswick, C., Clements, D. H, Sarama, J., Banse, H. W., and Day-Hess, C. A. (2019). Double impact: Mathematics and executive function. *Teaching Children Mathematics* 25(7): 416–426.

Klein, M. (2007). How is it that learning mathematics in the early years can become so difficult? A post-structuralist analysis. *Contemporary Issues in Early Childhood* 8(4): 313–319.

Klibanoff, R. S., Levine, S. C., Huttenlocher, J., Vasilyeva, M., and Hedges, L. V. (2006). Preschool children's mathematical knowledge: The effect of teacher 'Math Talk'. *Developmental Psychology* 42(1): 56–69.

Listening to Learn (2021). A K–5 digital interview tool. Available at: **www.listeningtolearn.com**

Malaguzzi, L. (1993). History, ideas and basic philosophy. In: C. Edwards, L. Gandini and G. Forman (eds.) *The Hundred Languages of Children*. Norwood: Ablex.

Mason, J. H., Burton, L., and Stacey, K. (1982). *Thinking Mathematically*. Harlow: Pearson Education.

Mason, J. H., Oliveria, H., and Boavida, A. M. (2012). Reasoning reasonably in mathematics. *Quadrante* XXI(2). Available at: **https://core.ac.uk/download/pdf/62692191.pdf**

Meinck, S., and Brese, F. (2019). Trends in gender gaps: Using 20 years of evidence from TIMSS. *Large-Scale Assessments in Education* 7(8). Available at: **https://link.springer.com/article/10.1186/s40536-019-0076-3**

Moss, J., Hawes, Z., Naqvi, S., and Caswell, B. (2015). Adapting Japanese lesson study to enhance the teaching and learning of geometry and spatial reasoning in early years classrooms: A case study. *ZDM Mathematics Education* 47: 377–390.

Moylett, H. (2018). Helping young children become great learners: Observing and supporting self-regulation. *Early Education Journal* 84: 7–9.

Mulligan, J., Kemp, M., and Highfield, K. (2008). *Encouraging Mathematical Thinking through Pattern and Structure: An Intervention in the First Year of Schooling*. Available at: **www.researchgate.net/publication/234657097**

National Numeracy. (2019). *Building a Numerate Nation: Confidence, Belief and Skills*. Lewes: National Numeracy. Available at: **www.nationalnumeracy.org.uk/sites/default/files/building_a_numerate_nation_report.pdf**

Noddings, N. (1984). *Caring: A Feminine Approach to Ethics and Moral Education*. London: University of California Press.

NRICH Primary Team. (2014). *Reasoning: Identifying Opportunities*. Available at: **https://nrich.maths.org/10990**

Nunes, T., Bryant, P., Sylva, K., and Barros, R. (2009). *Development of Maths Capabilities and Confidence in Primary School*. Research Report DCSF-RR118. Oxford: University of Oxford.

Nunes, T., Bryant, P., Evans, D., and Barros, R. (2015). Assessing quantitative reasoning in young children. *Mathematical Thinking and Learning* 17:2–3, 178–196. DOI: 10.1080/10986065.2015.1016815

Nursery World. (2007). *The One Hundred Languages of Children*. 21 March 2007. Available at: **www.nurseryworld.co.uk/news/article/the-one-hundred-languages-of-children**

OFSTED. (2013). *Mathematics in School Inspection: Information Pack for Training*. London: HMSO. Available at: **www.whatdotheyknow.com/request/154950/response/384078/ attach/15/Mathematics%20information%20pack.pdf?cookie_passthrough=1**

OFSTED. (2019a). *Early Years Inspection Handbook for Ofsted Registered Provision*. Reference 180040. Manchester: Crown Copyright.

OFSTED. (2019b). *Schools Inspection Handbook*. Manchester: Crown Copyright.

Paley, G. V. (2004). *A Child's Work. The Importance of Fantasy Play*. London: The University of Chicago Press.

Papic, M. M., Mulligan, J. T., Highfield, K., Mackay-Tempest, J., and Garrett, D. (2015). The impact of a patterns and early algebra program on children in transition to school in Australian Indigenous communities. In: B. Perry et al. (eds.) *Mathematics and Transition to School, Early Mathematics Learning and Development*. New York: Springer. DOI 10.1007/9789812872159_14

Pascal, C., Bertram, T., and Rouse, L. (2019). *Getting it Right in the Early Years Foundation Stage: A Review of the Evidence*. Watford: Early Education. Available at: **www.early-education.org.uk/sites/default/files/Getting%20it%20right%20in%20the%20 EYFS%20Literature%20Review.pdf**

Pinczes, E. J., and MacKain, B. (1993). *One Hundred Hungry Ants*. Boston: Houghton Mifflin.

Ribeiro, L. A., Casey, B., Dearing, E., Berg Nordahl, K., Aguiar, C., and Zachrisson, H. (2020). Early maternal spatial support for toddlers and math skills in second grade. *Journal of Cognition and Development* 21(2): 282–311.

Robertson, J. (2017). *Messy Maths: A Playful, Outdoor Approach for Early Years*. Carmarthen: Independent Thinking Press.

Robson, S. (2010). Self-regulation and metacognition in young children's self-initiated play and reflective dialogue. *International Journal of Early Years Education* 18(3): 227–241.

Rogers, S., and Evans, J. (2007). Rethinking role play in the reception class. *Educational Research* 42(2): 153–167.

Ross, P. (2011). *Is the Use of Role Play throughout the Primary School a Beneficial Strategy for Developing Children's Enjoyment, Competency and Understanding of Mathematics?* Submitted in partial fulfilment of the requirements for the degree of Master of Education, University of Exeter.

Russo, J. (2015). Teaching with challenging tasks: Two 'how many' problems? *Prime Number* 30(4).

Schmitt, S. A., Korucu, I., Napoli, A. R., Bryant, L. M., and Purpura, D. J. (2018). Using block play to enhance preschool children's mathematics and executive functioning: A randomised control trial. *Early Childhood Research Quarterly* 44: 181–191.

Serret, N., and Gripton, C. (eds.) (2020). *Purposeful Planning for Learning: Shaping Learning and Teaching in the Primary School*. ProQuest Ebook

Sharp, C. (2002). *School Starting Age: European Policy and Recent Research*. Paper presented at the LGA Seminar 'When Should Our Children Start School?', LGA Conference Centre, Smith Square, London, 1 November 2002. Available at: **www.nfer.ac.uk/media/1318/44414.pdf**

Shulman, L. S. (1986). Those who understand: Knowledge growth in teaching. *Educational Researcher* 15: 4–14.

Siemon, D., Barkatsas, T., and Seah, R. (eds.) (2019). *Researching and Using Progressions (Trajectories) in Mathematics Education*. Boston: Brill Sense.

Siraj-Blatchford, I., Sylva, K., Muttock, S., Gilden, R., and Bell, D. (2002). *Researching Effective Pedagogy in the Early Years (REPEY)*. Research Report 356. London: DfES.

Su, F. (2020). *Mathematics for Human Flourishing*. New Haven/London: Yale.

Tanner, H., and Jones, S. (2007). Using video-stimulated reflective dialogue to learn from children about their learning with and without ICT. *Technology, Pedagogy and Education* 16(3): 321–335.

Taylor, T. (2016). *A Beginner's Guide to Mantle of the Expert: A Transformative Approach to Education*. Norwich: Singular Publishing.

Thompson, I. (ed.) (2008). *Teaching and Learning Early Number*, 2nd ed. Maidenhead: OUP (first published 1997).

Thouless, H., Gifford, S., Moses, K., and James, R. (2020). Reasoning about patterns. *Mathematics Teaching* 271(April): 30–34.

TIMSS (Trends in International Mathematics and Science Study). (1999). *TIMSS Repeat 1999*. Available at: **http://timssandpirls.bc.edu/timss1999.html**

Twomey Fosnot, C., and Dolk, M. (2002). *Young Mathematicians at Work: Volume 3, Constructing Fractions, Decimals and Percents*. Portsmouth: Heinemann.

Van Oers, B. (2013). Is it play? Towards a reconceptualisation of role play from an activity theory perspective. *European Early Childhood Education Research Journal* 21(2): 185–198.

Venenciano, L., and Dougherty, B. (2014). Addressing priorities for elementary school mathematics. *For the Learning of Mathematics* 34(1): 18–24.

Verdine, B. N., Golinkoff, R. M., Hirsh-Pasek, K., and Newcombe, N. S. (2017). Links between spatial and mathematical skills across the preschool years. *Monographs of the Society for Research in Child Development*. Available at: **http://onlinelibrary.wiley.com/doi/10.1111/mono.v82.1/issuetoc**

Vygotsky, L. S. (1978). *Mind in Society: The Development of Higher Psychological Processes*. Cambridge, MA: Harvard University Press.

Wagner, B. J. (1999). *Dorothy Heathcote: Drama as a Learning Medium*. Portsmouth, NH: Heinemann.

Walls, F. (2009). *Mathematical Subjects: Children Talk About Their Mathematics Lives*. London: Springer.

Whitebread, D., and Coltman, P. (2017). Developing young children as self-regulated learners. In: J. Moyles, J. Georgeson and J. Payler (eds.) *Beginning Teaching: Beginning Learning: Early Years and Primary Education* (4th ed.). UK: McGraw-Hill Education; pp. 121–138.

Whitebread, D., Basilio, M., Kuvalja, M., and Verma, M. (2012). *The Importance of Play: A Report on the Value of Children's Play with a Series of Policy Recommendations*. Brussels: Toy Industries of Europe (TIE). Available at: **www.csap.cam.ac.uk/media/uploads/files/1/david-whitebread---importance-of-play-report.pdf**

Williams, H. J. (2006). *Let's Pretend Maths*. London: BEAM Education.

Williams, H. J. (2014). *The Relevance of Role Play to the Learning of Mathematics in the Primary Classroom*. PhD thesis, School of Education, University of Roehampton, London.

Winter, R. (1992). 'Mathophobia', Pythagoras and roller-skating. In: N. Nickson and S. Lerman (eds.) *The Social Context of Mathematics Education: Theory and Practice*. London: Southbank Press.

Worthington, M., and van Oers, B. (2016). Pretend play and the cultural foundations of mathematics. *European Early Childhood Education Research Journal* 24(1): 51–66.

WWC (What Works Clearing House). (2013). *Teaching Math to Young Children*. Available at: **https://ies.ed.gov/ncee/wwc/PracticeGuide/18**

PHOTO AND TABLE CREDITS

CHAPTER 1

Page	Photo	Credit
xii		William Gray. Taken at The Grove Primary School.
4	1.2	Esther O'Connor. Taken at The British School of Brussels.
7	1.4	Esther O'Connor. Taken at The British School of Brussels.
8	1.5	Esther O'Connor. Taken at The British School of Brussels.
10	1.7	Esther O'Connor. Taken at The British School of Brussels.
13	1.8	Esther O'Connor. Taken at The British School of Brussels.
14	1.9	Esther O'Connor. Taken at The British School of Brussels.
16	1.10	Esther O'Connor. Taken at The British School of Brussels.

CHAPTER 2

Page	Photo	Credit
18		Esther O'Connor. Taken at The British School of Brussels.
21	2.1	Esther O'Connor. Taken at The British School of Brussels.
22	2.2	Simon Gregg. Taken at the International School of Toulouse.
23	2.3	Simon Gregg. Taken at the International School of Toulouse.
26	2.7	Esther O'Connor. Taken at The British School of Brussels.
37	2.18	Esther O'Connor. Taken at The British School of Brussels.
42	2.22	William Gray. Taken at The Grove Primary School.
43	2.23	William Gray. Taken at The Grove Primary School.
43	2.24	William Gray. Taken at The Grove Primary School.
44	2.25	William Gray. Taken at The Grove Primary School.
46	2.26	Maeve Birdsall. Taken at Co-op Academy Oakwood.
49	2.30	Esther O'Connor. Taken at The British School of Brussels.
50	2.31	Maeve Birdsall. Taken at Co-op Academy Oakwood.
51	2.32	Esther O'Connor. Taken at The British School of Brussels.
53	2.33	Esther O'Connor. Taken at The British School of Brussels.

CHAPTER 3

Page	Photo	Credit
56		William Gray. Taken at The Grove Primary School.
58	3.1	Esther O'Connor. Taken at The British School of Brussels.
59	3.2	Esther O'Connor. Taken at The British School of Brussels.
66	3.8	William Gray. Taken at The Grove Primary School.
67	3.9	William Gray. Taken at The Grove Primary School.
68	3.10	William Gray. Taken at The Grove Primary School.
69	3.11	William Gray. Taken at The Grove Primary School.
71	3.12	Simon Gregg. Taken at the International School of Toulouse.
72	3.13	Simon Gregg. Taken at the International School of Toulouse.
73	3.14	Simon Gregg. Taken at the International School of Toulouse.
74	3.15	Maeve Birdsall. Taken at Co-op Academy Oakwood.
75	3.16	Esther O'Connor. Taken at The British School of Brussels.
76	3.18	Simon Gregg. Taken at the International School of Toulouse.
77	3.19	Simon Gregg. Taken at the International School of Toulouse.
77	3.20	Simon Gregg. Taken at the International School of Toulouse.
77	3.21	Simon Gregg. Taken at the International School of Toulouse.
78	3.22	Simon Gregg. Taken at the International School of Toulouse.
79	3.23	Simon Gregg. Taken at the International School of Toulouse.
79	3.24	Simon Gregg. Taken at the International School of Toulouse.
81	3.25	Maeve Birdsall. Taken at Co-op Academy Oakwood.
81	3.26	Esther O'Connor. Taken at The British School of Brussels.
82	3.27	Maeve Birdsall. Taken at Co-op Academy Oakwood.
83	3.28	Maeve Birdsall. Taken at Co-op Academy Oakwood.
84	3.29	Maeve Birdsall. Taken at Co-op Academy Oakwood.
85	3.30	Maeve Birdsall. Taken at Co-op Academy Oakwood.
87	3.31	Maeve Birdsall. Taken at Co-op Academy Oakwood.
89	3.32	Maeve Birdsall. Taken at Co-op Academy Oakwood.
90	3.33	Maeve Birdsall. Taken at Co-op Academy Oakwood.
92	3.35	Esther O'Connor. Taken at The British School of Brussels.
100	3.43	Esther O'Connor. Taken at The British School of Brussels.
100	3.44	Esther O'Connor. Taken at The British School of Brussels.
101	3.45	Esther O'Connor. Taken at The British School of Brussels.
102	3.36	Esther O'Connor. Taken at The British School of Brussels.
104	3.48	Esther O'Connor. Taken at The British School of Brussels.
105	3.49	Esther O'Connor. Taken at The British School of Brussels.
106	3.50	Esther O'Connor. Taken at The British School of Brussels.
109	3.53	Esther O'Connor. Taken at The British School of Brussels.

CHAPTER 4

Page	Photo	Credit
112		Esther O'Connor. Taken at The British School of Brussels.
115	4.1	Maeve Birdsall. Taken at Co-op Academy Oakwood.
118	4.4	Maeve Birdsall. Taken at Co-op Academy Oakwood.
119	4.5	William Gray. Taken at The Grove Primary School.
120	4.6	Esther O'Connor. Taken at The British School of Brussels.
122	4.8	Esther O'Connor. Taken at The British School of Brussels.
124	4.10	Esther O'Connor. Taken at The British School of Brussels.
125	4.11	Maeve Birdsall. Taken at Co-op Academy Oakwood.
130	4.16	William Gray. Taken at The Grove Primary School.
131	4.17	Esther O'Connor. Taken at The British School of Brussels.
136	4.22	Esther O'Connor. Taken at The British School of Brussels.
138	4.24	Esther O'Connor. Taken at The British School of Brussels.
141	4.25	Esther O'Connor. Taken at The British School of Brussels.
142	4.26	Esther O'Connor. Taken at The British School of Brussels.
142	4.27	Esther O'Connor. Taken at The British School of Brussels.
143	4.28	Maeve Birdsall. Taken at Co-op Academy Oakwood.
143	4.29	Esther O'Connor. Taken at The British School of Brussels.
144	4.30	Esther O'Connor. Taken at The British School of Brussels.
145	4.32	Esther O'Connor. Taken at The British School of Brussels.
146	4.33	Maeve Birdsall. Taken at Co-op Academy Oakwood.
148	4.2	Table with thanks to Dr Sue Gifford.
150	4.34	Julie Herlihy.
151	4.35	William Gray. Taken at The Grove Primary School.
152	4.36	Maeve Birdsall. Taken at Co-op Academy Oakwood.
157	4.39	Maeve Birdsall. Taken at Co-op Academy Oakwood.

CHAPTER 5

Page	Photo	Credit
160		Maeve Birdsall. Taken at Co-op Academy Oakwood.
163	5.1	William Gray. Taken at The Grove Primary School.
164	5.2	Esther O'Connor. Taken at The British School of Brussels.

All other images and figures are courtesy of the author, Helen J. Williams.

INDEX

Locators in **bold** refer to tables.